Seabed Politics

Barry Buzan

Seabed Politics

PRAEGER SPECIAL STUDIES IN INTERNATIONAL POLITICS AND GOVERNMENT

Praeger Publishers New York Washington London

Library of Congress Cataloging in Publication Data

Buzan, Barry.
Seabed politics.

(Praeger special studies in international politics and government)
Includes bibliographical references and index.
1. Ocean bottom (Maritime law). 2. Marine resources and state. I. Title.
JX4426.B89 341.7'62 75-36406
ISBN 0-275-22850-9

PRAEGER PUBLISHERS
111 Fourth Avenue, New York, N.Y. 10003, U.S.A.

Published in the United States of America in 1976
by Praeger Publishers, Inc.

Printed in the United States of America

to my parents

ACKNOWLEDGMENTS

This book, like most others, would have been impossible to produce if the author had not received assistance from many people. Space and time were provided by the Institute of International Relations at the University of British Columbia as part of its project "Canada and the International Management of the Oceans." Research funding, which enabled me to attend three sessions of the Third UN Conference on the Law of the Sea, was provided by the Donner Canadian Foundation.

More than 40 delegates to the conference shared information and opinions with me, and without the help of their knowledge and insight, this book would scarcely have been worth writing. Once I had committed myself to paper, a number of people gave invaluable advice on how I could present my ideas more clearly. My debt to Barbara Johnson and Dan Middlemiss is very large, both for their comments on various drafts and for the pleasure and inspiration of working with them at the Institute. Mark Zacher, the director of the Institute, gave me unstinting support, advice, and encouragement at all stages, and provided a stimulating and unrestrictive atmosphere in which to work. Don McRae of the University of British Columbia Law School helped me to understand the viewpoint of international lawyers, and Ed Miles of the University of Washington helped me to grasp the complexities of international negotiating and gave advice on the final draft. Maureen Gitta gave me the pleasure of seeing my chaotic pages transformed into a tidy typescript, and bore with me through the crises of an inevitably inconvenient deadline.

To all of these and others, I am grateful for advice both accepted and ignored. Responsibility for the work as it now stands, however, rests with the author.

CONTENTS

Chapter		Page
	The State of Seabed Politics at the End of the SBC Meetings	205
	Notes	207
9	THE THIRD UN CONFERENCE ON THE LAW OF THE SEA, 1973-74, PART I: CARACAS	211
	Developments Preceding the Caracas Session	211
	The Caracas Session	218
	Notes	238
10	THE THIRD UN CONFERENCE ON THE LAW OF THE SEA, 1974-75, PART II: GENEVA	242
	The Geneva Session	244
	Conclusions	265
	Notes	268
11	SUMMARY AND CONCLUSIONS	269
	Economic and Technical Factors	269
	State Action	272
	Actions within International Organizations	277
	Factors Shaping the Process of the Negotiations as a Whole	279
	Factors Shaping the Negotiations on the Continental Shelf	281
	Factors Shaping the Negotiations on the International Seabed Regime and Machinery	285
	Conclusions	294
	Notes	300
POSTSCRIPT		301
INDEX		303
ABOUT THE AUTHOR		312

LIST OF TABLES AND FIGURES

INTRODUCTION

SEABED POLITICS

The subject of this book is the political process by which one of the last remaining "unowned" areas of the planet is being brought under legal control. From the point at which interest in the seabed reached a sufficient level to stimulate demands for political and legal control over all or part of the area, three basic legal options for a seabed regime were available. First, the seabed could be left as an unsupervised international area like the high seas, useable by anyone but not subject to claim. Second, it could be treated like a land area, subject to claim, and thereby allowed to come under the jurisdiction of states. And third, it could be brought under active international jurisdiction and supervision as a common property of humanity. This third option was dependent on the existence of a suitable international organization, and a variety of other options could be devised by instituting the different types of regime in separate zones.

The book addresses three fundamental questions. Why and how did the issue of ownership and control of the seabed become an international problem? How was the problem solved in terms of the options outlined above? And, most important, what factors were responsible for the nature of the solution? The answers to these questions are sought by examining three distinct but related aspects of the political process. The first aspect is the development of economic and technological factors tending to raise the value of the seabed. The emphasis here is placed heavily on seabed mining (though fishing, military, and other uses do play a part) and on the link between economic interest and claims to control or own. The second aspect is the actions of states in relation to the seabed, particularly the making of national claims to seabed areas. This activity is related to the larger historical process by which coastal states (those states whose maritime interests are located predominantly off their own coasts) have emerged as challengers to the traditional dominance of the maritime states (those whose maritime interests are located predominantly away from their own coasts).

The third aspect is action within international organizations in relation to the seabed, with particular emphasis on the possibilities for such action resulting in an outcome involving active international jurisdiction and supervision. This is much the most complex of the three aspects, and the examination of it focuses on a number of related questions. First, how does an international organization acquire the policy initiative in relation to such an issue, and what is the relationship between its initiative and the claims and interests of states? Second, what kinds of power can competing alignments of states mobilize within an international negotiation? And third, in combination with these factors, does the nature of the international negotiating process itself preclude the radical developments in international organization necessary to implement a regime involving active international jurisdiction and supervision over valuable economic resources? Emphasis here is placed on factors that tend to slow down the negotiating process to such an extent that it becomes effectively unworkable as a means of taking action.

The first ten chapters of the book take the form of a chronological narrative tracing the linked development of economic and technological factors, national claims, and action within international organizations in relation to the seabed. This approach was chosen primarily to facilitate access to an important but increasingly complex contemporary issue for those readers new to it. Since most of the existing literature deals only with isolated segments of the subject, an understanding of the issue as a whole has been difficult to gain. In addition, much of the literature, which has been dominated by international lawyers, suffers from an inadequate historical perspective. The chronological approach offered a useful opportunity to reinterpret many events from the somewhat neglected point of view of the political analyst, concentrating on the movements of power underlying the evolution of international law. In the last chapter, the three aspects of the political process on which attention has been concentrated are linked together into a simple model of seabed politics, which is then used as a framework to consider the questions posed at the opening of the study. An attempt is also made to draw conclusions from the example of seabed politics about the implications of the international political process both for the international law that is its product, and for the instigation of radical advances in the structures, powers and functions of international organization by the use of carefully planned policy initiatives within the United Nations.

The numerous links between seabed and ocean politics pose something of a problem in that the emergence of a legal regime for the seabed is paralleled by the transformation of the legal regime of the oceans. On the whole, the physical differences between the seabed and the water column make it relatively easy to separate the politics of the two, and this separation is justified by the fact that ocean politics over such a time scale is a subject too large for any one book.[1] Difficulties do arise, however, in dealing with significant elements that are important to seabed politics but outside the scope of the present study. Examples of such elements are: the domestic politics of states, particularly the United States; other law of the sea issues, particularly seabed arms control, sedentary fisheries, and scientific research; and aspects of international politics such as the United Nations Committee on Trade and Development (UNCTAD), the Group of 77, and the various regional groups. The policy followed has been to outline these elements where they are relevant, but not to delve into them in detail.

For readers unfamiliar with the seabed, a brief description of its physical characteristics, its resources, and the terminology associated with it is included here. No attempt has been made to link this description to states by way of a systematic list of national maritime attributes such as coastline length, area of continental shelf, and so forth, because a comprehensive literature covering this already exists.[2]

THE SEABED

The seabed is a vast spread of submerged land covering over 360 million square kilometers of the earth's surface, an area 24 times greater than that of North America. Geographical characteristics divide it into two more or less distinct regions, the continental margin and the deep ocean floor. The margin is the land that is geologically associated with the continental mass. At present it constitutes approximately 15 percent of the seabed area, but this figure varies with the level of the oceans, and would be much less during an ice age. The margin itself can be divided into three regions: the continental shelf, which is a natural extension of the land, marked by a shallow inclination toward greater depths; the continental slope, which marks a break in the shelf and plunges at much steeper inclinations to great depth; and the continental rise, which is an accumulation of sediment at the base of the slope which tapers off, again at a shallow inclination, until it merges with the deep ocean floor.

The continental shelf may vary in width from a few hundred yards, as on some Pacific islands, to 800 miles, as off the coast of Siberia; the average width is around 40 miles. The break to the slope may occur at depths as shallow as 165 feet or as great as 1,650 feet, the average here falling between 435 and 660 feet. The slope is seldom more than 20 miles wide, and drops rapidly to the beginning of the rise at 6,600 to 9,900 feet. Together the shelf and slope, sometimes referred to jointly as the terrace, comprise about three-fourths of the area of the margin. The rise may drop to as much as 16,500 feet and extend as much as 600 miles from the slope before it becomes indistinguishable from the ocean floor. In most areas, the shelf-slope-rise pattern is fairly clear, but there are many anomalies that confound attempts to generalize. These include cases such as the west coast of South America, where the slope plummets directly into a deep trench, and the Blake Plateau off Florida, where the slope is interrupted part way down by a second level of "shelf".

The deep ocean floor can also be divided into geographical regions. It consists of relatively flat abyssal plains, deep trenches, undersea mountains, mountain ranges, and mid-ocean ridges, which form a continuous pattern of expanding crust in all the world's oceans. Most of this enormous area has been little explored, and while its general outlines are clear, most of its detail remains to be discovered.

The seabed contains a great wealth and variety of mineral resources, but the problem of obtaining access to them has meant that, for the most part, they have played little role in economic history. Only a few examples, such as the Cornish tin mines and the British and Chilean coal mines, stand out in the historical record. It is not until after World War II that seabed mining comes into importance with the development of the offshore oil industry.

Seabed mineral resources fall into five general categories.[3] First are the subsurface minerals, many of which are extensions of land deposits, and most of which occur within the limits of the margin. These include oil and gas, sulphur, coal, salt, tin, and potash. They can be mined by sinking a shaft into the seabed either from the shore or through an artificial island (as in the case of tin and coal) or by drilling into the seabed (as in the case of oil, gas, salt, and sulphur). Prospects for this type of mineral under the deep ocean floor are unclear, but problems of extraction are well beyond the reach of any economic technology.

Second are the placer deposits. These are mostly heavy metals and diamonds that have been concentrated by wave action on beaches or by the action of rivers carrying down particles from onshore deposits. Concentration occurs continuously on beaches and in river valleys, and deposits have been left behind in submerged beaches and valleys from eras when the ocean level was lower than at present. Mining is usually done by dredging through the overlying, lighter layers, and is only practicable in relatively shallow water. The minerals found in this category include gold, platinum, diamonds, tin, magnetite, zirconium, ilmenite, rutile, titanium, monazite, chromite, mercury, bismuth, antimony, and tantalum. Mining of this type has been going on since the early years of the 20th century;

some of the better known deposits are the diamonds off South Africa, tin off Indonesia, various heavy metals off Sri Lanka, and iron sands (magnetite) off Japan. The high costs of ocean mining restrict this activity to relatively valuable deposits, because land deposits, if available, are invariably cheaper. Many known deposits in this category are left unexploited for this reason.

Third are bulk commodity deposits of relatively low unit value. These include sand, gravel, and cement sands (seashells), and are usually mined by dredging in shallow coastal waters. Because of the large volume of production, these bulk commodity mines represent one of the most valuable segments of the hard mineral ocean mining industry. Other bulk commodity sources are found in deeper waters, but these have little economic potential. The most notable are glauconite, a source of potash and potassium found on the continental slopes off many states, and the red clays (alumina), calcereous oozes (calcium carbonate), and siliceous oozes that cover much of the deep ocean floor in enormous quantities. Plentiful sources of these are available on land, so there is no incentive to develop expensive extraction technology capable of recovering them from great depths.

Fourth are the metalliferous muds that occur in selected areas where oceanic rifts are active. The best known deposit is the Red Sea brine pool sediments, but others are suspected in the Gulf of Aden, the Gulf of Aqaba, the Gulf of California, and the East Pacific Ridge. The muds contain zinc, copper, lead, silver, tin, gold, iron, and manganese, and they may well prove amenable to exploitation in the near future. Not too much is known yet about this type of deposit; it is not clear whether it is a relatively rare occurrence or a fairly common characteristic of oceanic ridge activity.

Fifth are the various types of nodules found on the surface of the seabed. These include barite nodules, which contain a relatively poor grade of barium sulphate and are of little commercial interest; phosphorite nodules, which contain phosphate of use to fertilizer manufacturers; and ferromanganese nodules, which are of great contemporary interest as a source of copper, nickel, cobalt, and possibly also manganese, molybdenum, lead, zinc, zirconium, and other metals. Phosphorite nodules usually occur on the outer shelf and upper slope, and are found off the shores of many states. The phosphate they contain is of a poorer quality than that available from land deposits, and this means that commercial interest in them is relatively low. However, they may prove of value to countries such as India and Mexico, which are forced to import large quantities of phosphate at considerable cost. Manganese nodules occur very widely in the world's oceans, but their metal content varies considerably with location and depth. At the moment attention is focused on a large area of the Pacific stretching from the coast of Central America to 180 degrees longitude, and from the equator to 20 degrees north latitude. It is here, beneath waters averaging 13,000 to 16,500 feet in depth, that nodules containing commercially valuable concentrations of nickel, copper, and cobalt are found. Technologies for mining deep sea nodules are still in the early stages of development, and these will be examined in more detail in later chapters. Manganese nodules have been the center of great activity and expectations since the mid-1960s, and they may become the basis of a large-scale mining industry.

Seawater itself is also a resource of considerable importance. Extraction of salt from seawater has a history more than 4,000 years old, and bromine and magnesium have both been produced from this source for several decades. Pure water is an increasingly important product of the sea, and the techniques used to obtain it have given rise to hopes that boron, aluminum, lithium, fluorine, and uranium might also be extracted commercially. The lure of gold, which is one of the many metals present in vast quantities in seawater, has exercised many imaginations. The problem is, however, that while a huge absolute amount of metal is present in the oceans, it is present only in such extremely dilute solution that it constitutes a very low grade ore, and would not nearly justify the costs of processing. Since seawater is not a fixed resource, in the sense that seabed minerals are fixed, it is not subject to the kind of political problems associated with the seabed; consequently it will not form an important part of this study.

Much attention has been placed on the fact that placer deposits, metalliferous muds, and nodules are all in a process of continuous formation, and are thus a "self-renewing" resource.[4] Hopes have been raised on this basis that seabed minerals constitute an inexhaustible supply of many metals at a time when land based sources appear to be running out. This line of argument has caused particular excitement with regard to manganese nodules. While the process of continuous formation is indisputable, it should not be given as much importance as the extent of the resource currently existing. The process of renewal is immensely slow by human standards, and is very unlikely to have any impact on the kinds of exploiting activity currently envisaged.

This vast accumulation of resources lay undisturbed through thousands of years of human history. The following chapters will attempt to describe the brief but increasingly intense process by which they have been lifted from the depths of obscurity to the heights of international politics.

NOTES

1. A heroic and interesting attempt to do this has recently been undertaken by E. Miles, "The Dynamics of Global Ocean Politics," in Douglas Johnston (ed.), Marine Policy and the Coastal Community: Studies in the Social Sciences (London: Croom Helm, 1976). Another recent effort in this direction is J. S. Nye, "Ocean Rule Making From a World Politics Perspective," Ocean Development and International Law 3, no. 1 (1975), pp. 29-52.

2. See John K. Gamble, Global Marine Attributes (Cambridge: Ballinger, 1974); R. D. Hodgson, "Islands: Normal and Special Circumstances," in John K. Gamble and Giulio Pontecorvo (eds.), Law of the Sea: The Emerging Regime of the Oceans (Cambridge: Ballinger, 1973), pp. 137-200; U.S. Department of the Interior, Summary Petroleum and Selected Mineral Statistics for 120 Countries, Including Offshore Areas, Geological Survey Professional Paper 817, 1973; U.S. Department of State, Bureau of Intelligence and Research, Office of the Geographer, International Boundary Study Series A, no. 46, "Limits in the Seas: Theoretical Areal Allocations of Seabed to Coastal States," August 12, 1972; Lewis M. Alexander, "Indices of National Interest in the Oceans," Ocean Development and International Law Journal 1, no. 1 (Spring 1973), pp. 21-49; R. D. Hodgson and T. V. McIntyre, "Maritime Commerce in Selected Areas of High Concentration," in T. A. Clingan and L. M. Alexander (eds.), Hazards of Maritime Transit, (Cambridge: Ballinger, 1973), pp. 8-17.

3. There is a considerable literature on this subject. See Miller B. Spangler, New Technology and Marine Resource Development (New York: Praeger, 1970); Brenda Horsfield and Peter Stone, The Great Ocean Business (London: Hodder and Stoughton, 1972); Feng-hui Wang, Mineral Resources of the Sea, UN Doc. ST/ECA/125, 1970; John L. Mero, The Mineral Resources of the Sea, (New York: Elsevier, 1965); Lewis M. Alexander (ed.), The Law of the Sea: The Future of the Sea's Resources (Kingston: University of Rhode Island, 1968); T. Saunders English (ed.), Ocean Resources and Public Policy (Seattle: University of Washington, 1973); National Petroleum Council, Petroleum Resources Under the Ocean Floor (Washington, 1969); Vincent E. McKelvey and Frank F. Wang, World Subsea Mineral Resources: Preliminary Maps, U.S. Geological Survey, Department of the Interior, 1969; David R. Horn (ed.), Ferromanganese Deposits on the Ocean Floor, International Decade of Ocean Exploration (IDOE), National Science Foundation, Washington, 1972; and UN Docs. A/AC.138/36, 73, 87 and 90, and A/CONF.62/25 and 37.

4. John Mero emphasizes this point of view in ibid., p. 18.

Seabed Politics

CHAPTER

1

THE DEVELOPMENT OF SEABED POLITICS UNTIL 1958

SEABED POLITICS BEFORE WORLD WAR II[1]

The early history of seabed politics is inseparable from that of ocean politics. Except for tidal and other shallow areas, the entire seabed was beyond the reach of humanity, and consequently was of little interest to anyone. A few places containing pearls or edible shellfish were known and coveted, but for the most part interest centered on the surface of the sea. On the surface, men could sail vessels for the purpose of fishing, trade, piracy, or war, and then as now they argued about whether the sea was a common property free for the use of all of whether it was, like the land, appropriable by anyone who had the power to enforce a claim.

In classical times, the Greeks and the Romans treated the seas as res nullius (belonging to no one, and therefore open to claim), but some Roman thinkers like Gaius and Justinian were already developing the notion of res communis (belonging to everyone, and therefore open to use but not appropriation). State practice after the fall of classical civilization tended toward the res nullius interpretation, with states' claiming either specific jurisdiction or complete sovereignty over parts of the sea. As early as the ninth century, Byzantium claimed jurisdiction over fishery and salt resources, and by the 15th century quarantine regulations and limits were quite common. Venice claimed the Adriatic Sea, and various states laid claims to the Baltic, largely on the basis of local naval power. This process culminated in 1493-94, when Spain and Portugal divided up most of the world's oceans between themselves on the basis of a Bull pronounced by Pope Alexander VI.

Reaction against the Portuguese and Spanish claims became visible as English and Dutch naval power challenged their hegemony on the oceans. By the early years of the 17th century jurists in Spain itself were questioning the closed sea (mare clausum) doctrine, and in 1609 Hugo Grotius published his famous pamphlet Mare Liberum attacking the closed sea concept. The steadily increasing use of the sea by many countries was, in any case, making the extravagant closed sea claims of Spain and Portugal untenable. The high seas thus reverted in theory and in practice to an open sea (mare liberum) regime, while the closed sea concept (after John Selden's defense of English coastal fishing rights, Mare Clausum, published in 1635) retreated to a belt of "territorial sea" bordering a state's coast. By the end of the 17th century the high seas/territorial seas distinction was firmly established, and the marine league of approximately three miles was becoming the common limit of coastal state jurisdiction.

This compromise provided a stable legal regime for over 200 years. Except for the long established four-mile territorial sea claims of Norway, Sweden, and Iceland, and the six-mile claim made by Portugal in 1885, the three-mile territorial sea was almost universally recognized. It was a sufficient boundary to protect the interests of most coastal states, and since the activities of man had not as yet noticeably strained the seemingly inexhaustible resources of the ocean, freedom of the seas appeared to be an ideal regime.

The seabed passed almost unnoticed through these changes of regime. Its only importance was in places where its living resources were accessible, such as the pearl beds off Ceylon and Venezuela and in the Persian Gulf, the sponge fishing grounds off Tunisia, and the various places where oysters, sea cucumbers, and other edible sedentary species could be harvested. In many of these cases traditional rights of exploitation were built up over the centuries, and these rights constituted one of the few exceptions to the three-mile limit on coastal state jurisdiction. The British Colonial Act of 1811 was one of the first legal instruments to embody a claim to the living resources of the seabed beyond the territorial sea.

It was not until the middle of the 19th century, under the impact of a rapidly developing science and technology, that the seabed began to acquire an importance of its own. At that time the first deep sea surveys were beginning to plumb the depths, and the laying of telegraph cables was making the seabed important. A cable was successfully laid across the English Channel in 1851, and though a trans-Atlantic cable failed in 1858, one was completed successfully in 1866. This activity had reached sufficient importance that a Convention for the Protection of Submarine Cables was held in Paris in 1884, and the boom in cable laying continued up to the end of the century. In addition, H.M.S. Challenger undertook one of the first great oceanographic expeditions from December 1872 to May 1876. In her world-circling tour, the Challenger added numerous soundings to the new maps of the deep seabed, and brought back many bottom samples, among which were found the first specimens of manganese and phosphorite nodules.

Limitations of the contemporary technology, and the availability of relatively plentiful land resources, however, caused commercial interest in seabed mineral resources to lag considerably behind scientific interest. Commercial mining of the seabed was limited to undersea extensions of shafts sunk from the land, mostly for coal, right up to the end of the 19th century. Several countries followed the example of Britain's Cornwall Submarine Mines Act (1858), which claimed such mines as crown territory, and nobody presented a significant challenge to such mining even when it extended beyond three miles. In the last years of the 19th century the oil industry made its first steps toward offshore drilling by extending piers out from beaches of California oilfields. But except where sites were adjacent to existing oilfields, as in California and Lake Maracaibo, there was little incentive for the industry to indulge in expensive offshore operations. Dredging for placer deposits in shallow waters also began early in the 20th century, but in general resources were so plentiful on land that the more costly offshore operations were not justified.

Despite the lack of commercial interest in the seabed, there were many technological developments that multiplied the possibilities of using this previously useless area. The Frasch process for liquifying sulfur, for example, was introduced in the early part of this century, and opened up the possibility of mining sulphur by using drilling techniques similar to those used for oil. Perhaps most important were the many advances made in military technology. Efficient submarines were being made by the turn of the century, and bottom-moored mines achieved prominence in the Russo-Japanese War of 1904-05 and in World War I, which also greatly stimulated research into the use of sound to locate submarines under water. This work led directly to the development of echo-sounding devices (sonar), which were enormously useful to oceanographers since they simplified the job of mapping the seabed by replacing the time-consuming lead and line as a means of taking depth measures.

Although these developments were modest by current standards, they were sufficient to attract attention to the seabed as an area apart from the ocean waters, and to establish it as a factor in the reviving debate over the limits of coastal state jurisdiction. The long and tedious job of mapping the seabed had, by the end of the 19th century, produced sufficient data to suggest that the continental shelf was a natural prolongation of the continental land mass. In 1910 Portugal used this natural extension argument to justify a claim to jurisdiction over fishing as far out as the 100-fathom depth line, and Russia based a 1916 claim to Arctic islands on the fact that the islands in question were linked to the mainland by the vast Siberian continental shelf. In addition, writers in Argentina, Spain, Cuba and elsewhere argued for extensions of coastal state jurisdiction based on the limit of the continental shelf.

These arguments, however, were not of much importance in the overall debate on the limits of national jurisdiction in the sea. An increasing number of states were becoming dissatisfied with the three-mile limit, but their concern was primarily for coastal fish stocks. Russia extended her jurisdiction to 12 miles in 1909 for fishing and customs purposes, but encountered strong opposition to this move from the traditional maritime powers. Unilateral extensions beyond the traditional norm raised the thorny questions of whether states had either the authority or the power to impose their jurisdiction over areas established by long usage as high seas, and whether custom had not created international norms that could be altered only by international agreement.

It was at least partly to settle this question that the League of Nations convened a Conference on the Progressive Codification of International Law in the spring of 1930. Seabed issues played almost no part in this conference, and a report on the "Exploitation of the Products of the Sea" prepared in 1926 as a document for the conference dealt entirely with living resources and made no mention of seabed or seawater minerals. The conference concerned itself primarily with the width of the territorial sea, and it is interesting here as a milestone in the breakdown of the traditional regime under the pressure to expand the limits of coastal state jurisdiction. Forty-two states attended (with the Soviet Union as an additional observer), and after extensive debate were completely unable to agree on a standard width for the territorial sea. Much of the disagreement concerned rights to coastal fish stocks.[2] The seabed was affected only because extensions of the territorial sea involved extensions of coastal state jurisdiction over the underlying continental shelf. By the end of the conference the positions of 36 states had become clear. These states were divided into four groups on the basis of the maximum limit they would accept for the territorial sea. They were further divided within those categories according to whether or not they wanted a contiguous zone beyond the territorial sea. Some refused to consider any such zone, some wanted it only for customs and sanitation functions, and some wanted it for security and fishery jurisdiction as well.

TABLE 1.1

Positions of 36 States on the Territorial Sea and Contiguous Zone, 1930

3 Miles	3 Miles and Contiguous Zone	4 Miles	6 Miles	6 Miles and Contiguous Zone	12 Miles
Australia	Belgium	Finland	Brazil	Cuba	Portugal
Canada	Egypt	Iceland	Chile	Iran	Soviet Union
China	Estonia	Norway	Italy	Latvia	
Greece	France	Sweden	Romania	Spain	
India	Germany		Uruguay	Turkey	
Japan	Poland		Colombia		
Netherlands			Yugoslavia		
Ireland					
South Africa					
United Kingdom					
United States					
Denmark					

Note: Contained in the following sources is some disagreement about whether Chile should be categorized under 3 or 6 miles, Iceland under 3 and Contiguous Zone or 4 miles, Portugal under 6 or 12 miles, and Brazil under 3 or 6 miles. In the listing given here, majority opinion has prevailed.

Source: Garcia Amador, The Exploitation and Conservation of the Resources of the Sea (Leyden: Sythoff, 1963), p. 29; Edward Miles, "Technology, Ocean Management, and the Law of the Sea: Some Current History," Denver Law Journal 46, no. 2 (1969), pp. 245-46; C. J. Colombos, The International Law of the Sea (London: Longmans, 1967), p. 104; and Inter-American Juridical Committee, Opinion on the Breadth of the Territorial Sea, OEA/Ser. I/VI.2 CIJ-80 (Washington, D.C.: Pan American Union, 1966): 22.

These alignments are interesting because they illustrate the extent to which dissatisfaction with the old norms of coastal state jurisdiction had developed. One-third of the states at the conference clearly favored a limit outside the prevailing norm, and nearly all of these were less developed states that could not take much advantage of the opportunities provided by a worldwide freedom of the seas regime. The conference was quite clearly divided between the maritime states and their allies on the one side, and the smaller and nonmaritime coastal states on the other. The maritime states were interested in maintaining a system that had grown out of their shipping, fishing, and naval practice, while the smaller and nonmaritime states were interested in altering the balance of a system that gave them no advantages and in some cases, especially coastal fishing, worked visibly against their interests.

Following the failure of the conference to agree on a limit for the territorial sea, a steady trickle of states, many of them Latin American, unilaterally increased their territorial limits beyond the three-to-four-mile norm. Uruguay (1930), Colombia (1930), Iran (1934), Cuba (1934), Greece (1936), and Italy (1942) all declared six-mile limits; Mexico (1944) declared nine miles; and Guatemala (1940) and Venezuela (1941) declared 12 miles.

These claims marked the beginning of a long process of unilateral extension of coastal state jurisdiction that was to undermine the old regime of narrow coastal state jurisdiction. This process was fueled by a rapid increase in the variety and intensity of uses of the ocean, an increase that swept away the assumption of inexhaustibility of resources on which the old regime had been based. It was to involve massive transfers of resources from free use areas to coastal state jurisdiction, and though it began with quite modest territorial sea claims, it spread quickly to include much more ambitious claims.

THE FIRST NATIONAL CLAIMS TO THE SEABED, 1945-50

Immediately following World War II, there was a noticeable quickening in the tempo of expanding coastal state jurisdiction. States with an interest in extending their marine boundaries could do so in several ways other than a simple increase in the measured breadth of the territorial sea. They could, for example, change the baseline from which the territorial sea was measured by moving it from the high-water to the low-water mark, or they could institute straight baselines across bays and inlets, thereby increasing the area of their internal waters. They could also declare special contiguous zones adjacent to the territorial sea in which they would exercise jurisdiction only in relation to specified functions. These zones might cover activities relating to fishing, customs and health, pollution, traffic control systems, security, or seabed mining. States could also extend their jurisdiction by claiming all or part of the continental margin as a part of their sovereign national territory, without that claim affecting the status of the superjacent waters. In the period 1945-50, claims to sovereignty or jurisdiction over the seabed replaced territorial sea claims as the most popular way to extend national maritime boundaries.

The trend toward seabed claims started during World War II. Venezuela and Britain signed the Treaty of Paria in February 1942, dividing the seabed in the Gulf of Paria between Venezuela and Trinidad; in January 1944 Argentina made a somewhat vague claim to the resources of its continental shelf. While these were precedent-setting claims, they do not appear, in the general furor of the war, to have made much impact. It was not until September 28, 1945, when President Truman claimed "the natural resources of the subsoil and seabed of the continental shelf beneath the high seas but contiguous to the coasts of the United States as appertaining to the United States [and] subject to its jurisdiction and control," that claims to the continental shelf became a major element in the extension of coastal state jurisdiction.

The Truman Proclamation had roots going back as far as 1937, when President Franklin D. Roosevelt was thinking about a unilateral fishing claim related to the continental shelf off Alaska. It became a serious proposition in June 1943, when an interdepartmental committee was established by the Departments of State, Justice, and Interior to look into the problems of acquiring continental shelf and coastal fishery resources. The enormous drain on American resources caused by the war, and particularly the lavish export of oil, had stimulated an interest in continental shelf resources that otherwise would probably not have developed so quickly. The offshore oil industry was still developing relatively slowly at the time, and did not establish a well beyond the three-mile limit until 1947. A paper proposing exclusive reservation of continental shelf resources contiguous to the United States was circulated in July 1944, and after much revision, the final proclamation was ready by the late summer of 1945.[3]

The Truman Proclamation was a landmark in seabed politics. Not only did it reveal that advancing technology was bringing seabed mineral resources within the realm of economic accessibility, but it was also the first claim by a major maritime power to jurisdiction over the continental shelf beyond the territorial sea. The proclamation was carefully worded, making claim neither to the waters superjacent to the continental shelf nor to the continental shelf itself. It scrupulously avoided any mention of sovereignty, and claimed jurisdiction only over resources. It was thus explicitly a functional, not a territorial, claim. The proclamation itself contained no definition of the continental shelf, but a White House press release issued simultaneously defined it as an area of about 750,000 square miles contiguous to the continent and covered by no more than 600 feet of water.

The Truman Proclamation was a conceptual breakthrough. It gave great weight to the principle stated in its preamble that the mere existence of the continental shelf as "an extension of the land-mass of the coastal nation and thus naturally appurtenant to it" was sufficient basis for a claim by the adjacent coastal state. This was a huge step beyond the old norm, under which only the long usage of seabed fisheries was considered legitimate justification for such a claim. The norm suggested by the proclamation was based instead on physical adjacency and anticipation of use.

Despite its radical nature, the Truman Proclamation encountered no open opposition from any other state, and this greatly increased its validity as a precedent. It is important to note, however, that the American government made it clear to other governments that it had no intention of reserving continental shelf resources exclusively for its own nationals, any more than it excluded foreigners from participation in exploitation of American land resources. The government's primary concern was to "assert the necessary control over . . . operations off the coasts of the United States to guard against the depletion of . . . mineral resources and to regulate, from point of view of security, the activities of foreigners in proximity to [the] coast."[4]

By giving the principle of geologically-related adjacency some legitimacy, the Truman Proclamation paved the way for dozens of other claims to the continental shelf. In the years up to 1950, 30 unilateral claims followed the American initiative. Many Latin American countries were quick to take advantage of the opportunity, including Mexico (1945), Argentina (1946), Panama (1946), Chile (1947), Peru (1947), Costa Rica (1948), Guatemala (1949), Brazil (1950), Nicaragua (1950), Honduras (1950), and Ecuador (1950). Britain made no claim for itself, but did so on behalf of Trinidad (1945), Jamaica (1948), nine Persian Gulf sheikdoms (1949), and some of its other colonies including the Bahamas, British Honduras, and the Falkland Islands. Iceland (1948), Saudi Arabia (1949), Iran (1949), the Philippines (1949), and Pakistan (1950) also made claims to the continental shelf in this period. By comparison, only three states in those years chose to extend their maritime jurisdiction by widening their territorial seas: Yugoslavia to six miles (1948), Ecuador to 12 miles (1950), and El Salvador to 200 miles (1950). The Truman Proclamation, then, marked both a change in the nature, and an acceleration, of the trend to increasing coastal state jurisdiction established after the 1950 conference.

But although the Truman Proclamation was the trigger that set off this round of unilateral claims, it did not become the model on which most of them were based. Britain followed the American pattern in claiming only functional jurisdiction over resources, and so did Mexico, Guatemala, Iran, and the Philippines. Most of the Latin American countries, however, followed the lead of Argentina and claimed the shelf as part of their sovereign territory. In addition, Iceland and eight Latin American countries explicitly linked the continental shelf to the fish resources in the superjacent waters, and so used the logic of the continental shelf claim to extend their jurisdiction over coastal fisheries. This link between the shelf and the superjacent waters had been developed in the Argentine claim of 1946 to sovereignty over the shelf and the "epicontinental" sea. Chile, Costa Rica, Ecuador, El Salvador, Honduras, Panama, and Peru all followed the Argentine lead. The Argentine claim, like most others of this period, made little attempt to define the limits of the continental shelf. Only Mexico, Brazil, Ecuador, Nicaragua, and Pakistan followed the U.S. lead by setting limits at the 600-feet/200-meter depth line.

It was, however, the much more specific and far-reaching claim of Chile that had the greatest international impact. Motivated in part by the activity of foreign fishing and whaling fleets off its coasts, the Chilean government followed the Argentine model and proclaimed

> its national sovereignty over all the continental shelf adjacent to the continental and island coasts of its national territory, whatever may be their depth below the sea . . . [and] over the seas adjacent to its coasts whatever may be their depths, and within those limits necessary in order to reserve, protect and exploit the natural resources of whatever nature found on, within and below the said seas.

By way of making this claim operational it declared "protection and control . . . over all the seas contained within the perimeter formed by the coast and the mathematical parallel projected into the sea at a distance of 200 nautical miles from the coasts of Chilean territory."[5] Peru followed the Chilean example by enacting similar legislation, Costa Rica and Honduras produced somewhat weaker versions

(Honduras seems to have abandoned the 200-mile idea shortly after proclaiming it), and El Salvador carried the argument to its logical extreme and declared a full 200-mile territorial sea. Many of these more extreme Latin American claims, particularly those of Argentina, Chile, Ecuador, Peru, and El Salvador, met with vigorous protest from the United States, Britain, and others. These protests, like the claims themselves, were motivated more by fishing and security concerns than by seabed interests. They can be interpreted as attempts to stop the use of seabed claims, which were seen by the maritime powers as legitimate (barring the question of sovereignty), as a springboard for making much wider fishing and territorial sea claims, which the maritime powers opposed.

The result of this first round of unilateral claims to the continental shelf was, then, a mixed bag of legislation concentrated largely in the Western Hemisphere and around the Persian Gulf. The claims were so varied and contested that when a case concerning continental shelf rights finally came before international arbitration in 1949, the arbitrator decided that the "continental shelf doctrine" was not yet an established rule of international law.[6] While it was clear that many of the claims were modeled on previous claims, there is little evidence to suggest that any significant regional cooperation explains their geographical concentrations.

In the case of the Persian Gulf states, and also of the United States and Trinidad, the motivating interest was oil. The shallow bed of the Persian Gulf was known to contain extensions of the rich onshore oil deposits, and the American experience off California proved that these were, or soon would be, accessible. The same logic applied to Trinidad in the Gulf of Paria.

In the case of Latin America, where the regional concentration was exceptionally strong, the explanation was more complex. By 1950, 15 of the 17 Latin American countries had made some sort of claim beyond the three-mile limit: only Haiti and the Dominican Republic did not have such legislation. These claims varied widely, from the six- and 12-mile territorial sea limits (several of which dated to before 1945), through the miscellany of continental shelf claims, to the 200-mile zone and 200-mile territorial sea claims. The Organization of American States (OAS), founded in 1948, continued a long tradition of continental meetings to discuss issues of common concern, but it does not appear to have played more than an informal communication role in stimulating or coordinating the Latin American claims at this stage. With the exception of Venezuela, oil was not an immediate prospect for any of the Latin American countries, and not even a good long-range prospect for most. Chile mined coal under the seabed from land shafts, but other than this uncontested activity, none of the Latin American countries had real or prospective interests in hard minerals from the seabed.

The primary motive for the Latin American countries to increase their maritime jurisdiction appears to have been concern over actual and anticipated foreign fishing off their coasts.[7] This concern existed in spite of the facts that the fish catch off Latin America amounted to less than 3 percent of the world total by weight (1956), and that only Peru, Chile and Brazil ranked among the top 30 fishing states (respectively 23, 28, and 29).[8] Despite these figures, fish were seen locally as a resource capable of immediate exploitation with good prospects for a substantial increase in the total catch.

While the seabed itself was of little practical use to most Latin American countries, it provided a relatively firm basis on which to build unusually broad fishing claims. There was no precedent for the direct claim to control over fishing beyond 12 miles either by fishing zones or by declaration of a territorial sea. But given the logical wedge of the Truman Proclamation, and by changing jurisdiction to sovereignty, it was possible to argue for extended fishing zones based on the quite real link between fishery resources and the continental shelf.[9] From that point it was but a short jump to the 200-mile zone of Chile and the 200-mile territorial sea of El Salvador.

This process was aided by five factors: first, by the vigorous tradition of nationalism in Latin America, which gave a highly positive value to sovereignty; second, by the geographical fact that nearly all the Latin American states border on the open ocean and therefore do not have delimitation problems with opposite states; third, by the circumstance that the continent had not been much involved in the war, and was the only group of "old" independent states not preoccupied with recovering from that conflict; fourth, by the particular circumstances of Chile, Ecuador, and Peru, whose very narrow continental shelves offered them almost no gain from the new continental shelf doctrine, and who were therefore motivated to develop a doctrine that would compensate them for this disadvantage; and fifth, by the triggering process started by the Truman Proclamation, which meant that each new national claim increased the incentives for other states to jump on the bandwagon.

On this basis, and at first in a largely uncoordinated manner, the Latin American states emerged as the leading exponents of expanding coastal state jurisdiction. Despite their lack of interest in the seabed itself, their use of it to promote other interests cast them in a major role in the more specific and rapidly developing drama of seabed politics.

THE FIRST REGIONAL AND INTERNATIONAL RESPONSES TO NATIONAL CLAIMS, 1951-58

The burst of unilateral continental shelf claims and protests that followed the Truman Proclamation, and the steady trickle of unilateral extensions of the territorial sea that had been going on since 1930, finally provoked responses on both regional and international levels. On the regional level, the Inter-American Council of Jurists (IACJ), which was created as part of the Organization of American States (OAS) in 1948, took up the challenge offered by the numerous maritime claims of its members.[10] At its first meeting in June 1950, the IACJ discussed the question of ocean boundaries and decided to request its Juridical Committee (IAJC) to prepare a study on territorial waters and related questions. On the international level, the International Law Commission (ILC) at its first session in 1949 chose the law of the sea as one of three subjects to which it would give immediate priority. The ILC had been created by the United Nations in 1947 to undertake the codification and progressive development of international law by preparing draft conventions which would be submitted to international conferences for approval. Thus the IACJ directed itself toward finding common elements in the multilateral and unilateral acts of the American states, while the ILC attempted to find, or create, sufficient common ground to build a universally accepted, uniform, international law of the sea. Both groups completed the bulk of their work on this subject between 1950 and 1956, but their activity did not stem the flow of national claims, and in some cases even stimulated it.

The work of the ILC led directly to the First United Nations Conference on the Law of the Sea in 1958, but in the years between 1951 and that conference 28 more states unilaterally extended their maritime jurisdiction. Most of these were new claims, but some were additions to, or clarifications of, previous claims to the continental shelf or to the territorial sea beyond three miles. Seventeen states made claims to the continental shelf, and as before, these claims varied greatly in form. Many used vague geographical terms, some specified a depth limit, three picked up definitions being worked out by the ILC and the IACJ, and one, Ecuador, followed in the 200-mile footsteps of Chile and Peru. In addition, there continued to be a major division between those claiming sovereignty over the shelf and those claiming merely jurisdiction over its resources. Latin American states remained prominent in the list of claimants—the Dominican Republic (1952), Ecuador (1952), Venezuela (1953 and 1956), Guyana (by the United Kingdom, 1954) and Guatemala (1949 and 1956)—but states from most other regions of the world joined the ranks in greater numbers. Claims to the continental shelf were made in Asia by South Korea (1952), India (1955), Iran (1949 and 1955), Burma (1957), Cambodia (1957), and Sri Lanka (1957); among the Arab states by Iraq (1957) and Egypt (1958); and among the Western states by Israel (1952), Australia (1953), Portugal (1956), and Spain (1958).

In this period there was also a substantial revival of claims to a territorial sea beyond three miles. Eighteen states made such claims, and as with the continental shelf, there was little uniformity in the declared limits. These claims were only partly the result of the fact that newly independent countries were entering the international system. At this stage the flow of new states had not yet reached the major proportions it was to assume after 1960. Most of the claims came from long-established states, and showed a pronounced tendency toward the 12-mile limit pioneered by the Soviet Union. No particular regional trend emerged from these claims, except that no Western states had yet overcome the narrow limits tradition sufficiently to claim 12 miles. A good number of states had claims to both the continental shelf and a wider territorial sea, thereby reinforcing the trend started in Latin America to extend coastal state jurisdiction both on the seabed and in the water column. Extensions of national jurisdiction in the territorial sea were also assisted by a decision of the International Court of Justice in 1951 to allow Norway the use of straight baselines, along the heavily indented parts of its coast, from which to measure its territorial sea.

The effect of this growing weight of unilateralism was to raise the pressure on those working to find a uniform basis for the law of the sea, and, increasingly, to define the terms by which they could seek universal solutions. Balancing this, however, was the inevitable increase in disagreements and disputes that accompanied unilateral claims. In addition to the protests raised by Latin American actions, Japan protested Australia's continental shelf claim; Australia protested Indonesia's claim to a 12-mile territorial sea around the outer edge of the Indonesian archipelago; and Britain and Iceland embarked on the first rounds of the cod war based on fishing limits arising from Iceland's claim to the continental shelf. These disputes pointed out the limitations of unilateralism in an area where so many interwoven interests were at stake, and so strengthened the incentives for states to support regional and international initiatives.

One other development related to unilateralism took place in this period, and that was the promulgation on subregional levels of declarations by groups of states with common maritime policies. Naturally enough, this activity started in Latin America, where the process of unilateral maritime claims was most advanced, and where such claims had provoked the most strenuous opposition.

TABLE 1.2

Claims of 18 States to the Territorial Sea, 1951-58

12 Miles	10 Miles	6 Miles	5 Miles	4 Miles
Bulgaria (1951)	Albania (1952)	India (1956)	Cambodia (1957)	Finland (1956)
Romania (1951)		Israel (1956)		
Ethiopia (1953)		Spain (1957)		
Libya (1954)		Sri Lanka (1957)		
Indonesia (1957)		Thailand (1958)		
Panama (1958)				
Iraq (1958)				
Saudi Arabia (1958)				
Egypt (1958)				
China (1958)				

Source: Data derived from a wide variety of sources, the most comprehensive of which is U.S. Department of State, Office of the Geographer, Bureau of Intelligence and Research, Limits in the Seas No. 36: National Claims to Maritime Jurisdiction, April 1974.

The first and most famous of these declarations was that made in Santiago in August 1952 by Chile, Ecuador, and Peru. The Santiago Declaration paralleled closely the earlier legislation of Chile and Peru, but represented a new claim for Ecuador, largely superceding its 1950 claim to a 12-mile territorial sea and a continental shelf out to 200 meters. It stated that each of the three governments possessed "sole sovereignty and jurisdiction over the area of sea adjacent to the coast of its own country and extending not less than 200 nautical miles from the said coast." This included "sole sovereignty over the sea floor and subsoil" but did not disallow "the necessary restrictions on the exercise of sovereignty and jurisdiction imposed by international law to permit the innocent and inoffensive passage of vessels of all nations through the zone."[11] Since the claim was directly to the sea and the seabed, and not just to resources, the Santiago Declaration looked very much like a 200-mile territorial sea claim by the three governments. They preferred not to follow El Salvador in this matter, however, and insistently referred to their creation as "the maritime zone." Costa Rica, which had similar 200-mile zone legislation, adhered to the declaration in October 1955, but never ratified it; she eventually withdrew in 1966. Chile, Ecuador, and Peru signed an additional agreement at Lima in December 1954, in which they made provision for the common defence of their zones.

The Santiago Declaration formed the basis of a long-standing and active alliance in defence of the most extreme coastal state position then existing. While El Salvador's 200-mile territorial sea claim had set a precedent, in practice it amounted to only about 2.2 percent of the area claimed by Chile, Ecuador, and Peru.[12] These three states successfully established and maintained a 200-mile zone along the coast of half a continent. In so doing, they kept alive through a long period of opposition and even ridicule an idea that eventually became a norm of international practice.

The other example of this type of activity was the Declaration of Antigua Guatemala signed by Guatemala, El Salvador, Honduras, Nicaragua, and Costa Rica at the founding meeting of the Organization of Central American States (ODECA) in August 1955. After a stormy series of meetings that went on despite a minor armed conflict between Costa Rica and Nicaragua, the conference managed to approve 13 resolutions on miscellaneous subjects. One of these was a claim to defend the "patrimony" of the Central American states, which was defined as including the continental shelf and epicontinental sea. Since all of the signatories already had claims to the continental shelf, the declaration did not attract much attention, but it is interesting as the origin of the term "patrimonial sea," which was to become important a decade and a half later.[13]

The Regional Response: Latin America

In response to the IACJ's request of June 1950, the Juridical Committee produced five draft articles constituting a Draft Convention on Territorial Waters and Related Questions. This work was completed in July 1952, less than one month before the Declaration of Santiago, and was adopted in the committee by a close vote of 4 to 3, with the United States, Colombia, and Brazil opposing.[14] The committee proposed that:

> 1. The signatory States recognize that present international law grants a littoral nation exclusive sovereignty over the soil, subsoil, and waters of its continental shelf, and the air space and stratosphere above it, and that this exclusive sovereignty is exercised with no requirement of real or virtual occupation.

> 2. The signatory States likewise recognize the right of each of them to establish an area of protection, control, and economic exploitation, to a distance of two hundred nautical miles from the low-water mark along its coasts and those of its island possessions, within which they may individually exercise military, administrative, and fiscal supervision over their respective territorial jurisdictions.[15]

These two articles reflected the influence of Argentina's epicontinental sea and Chile's maritime zone, and the Draft Convention as a whole resembled the shortly forthcoming Declaration of Santiago. The committee appeared to take the position that the best way to achieve a uniform law was to accept the largest existing unilateral claims as its norm, although it did recognize the existence of disagreement on this and the need for further study of the subject.

The Draft Convention was submitted to the second meeting of the IACJ in April 1953, where it became the basis of a lengthy but indecisive discussion. The IACJ decided not, at that time, to express an opinion on "the nature and scope of claims that riparian States may make to their continental and island shelves, and to their territorial waters." Instead, it recognized the strong differences of opinion that existed among its members, and sent the subject back to its Juridical Committee for further study. It did, however, go so far as to approve, over the objection of the United States, this statement proposed by Cuba and Mexico:

> It is an obvious fact that development of technical methods for exploring and exploiting the riches of these zones has had as a consequence the recognition by international law of the right of such States to protect, conserve, and promote these riches, as well as to ensure for themselves the use and benefit thereof.[16]

This statement was a concise version of the argument that underlay many Latin American claims. It was a recognition that technological developments, especially those related to fishing and whaling, were rendering obsolete the long-accepted principle that ocean resources were inexhaustible. Further, it was a recognition that the Latin American countries were very far down in the hierarchy of technological capability, and that consequently it was not they, but others, who were benefiting from this change. Since resources were now considered capable of depletion, and since newly accessible mineral resources in the continental shelf were finite by definition, the only way in which technologically less-developed countries could maintain or advance their interests was by extending their jurisdiction over the resources themselves.

The subject of maritime boundaries came up again at the tenth Inter-American Conference of the OAS in 1954. Ecuador created a minor crisis in the proceedings by proposing a strong resolution along the lines of the Declaration of Santiago, but the conference was neither prepared nor willing to tackle this challenge. Instead, it agreed on a rather bland statement of interest in the whole question of maritime jurisdiction, and directed the Council of the OAS to convene a specialized conference on the subject. The Juridical Committee suspended its work in this area while awaiting the outcome of the specialized conference, but in January 1955 the OAS Council instructed the IACJ to consider the subject at its third meeting as part of the preparation for the conference.

The IACJ held its third meeting in Mexico City early in 1956, and after a long debate produced "The Principles of Mexico on the Juridical Regime of the Sea," a document it characterized as an "expression of the juridical conscience of the Continent." The principles included a denunciation of the three-mile territorial waters rule; a statement supporting unspecified extensions beyond that limit based on the needs of the coastal state; a reaffirmation of the rights of the coastal state to the seabed, subsoil, and resources of its continental shelf; and a statement supporting greatly increased coastal state rights over conservation and exploitation of coastal fisheries regardless of territorial sea limits.[17] The Principles of Mexico were passed by a vote of 15 to 1 to 5,[18] the United States opposing, and the Dominican Republic, Colombia, Cuba, Bolivia, and Nicaragua abstaining. The vote was a big defeat for the United States' three-mile position, and the American delegate denounced the principles as contrary to international law and interest, inadequately researched and prepared, politically motivated, and exceeding the competence of the IACJ. His opposition, however, merely highlighted the emerging, though still very vague, consensus among most of the Latin American countries on the question of maritime jurisdiction. The principles were submitted to the Specialized Conference on Conservation of Natural Resources: the Continental Shelf and Marine Waters, held less than two months later in Ciudad Trujillo.

The Ciudad Trujillo Conference was the last Latin American effort on the regional level before attention turned to the upcoming United Nations Conference on the Law of the Sea scheduled for 1958. By the time of the Ciudad Trujillo meeting, the International Law Commission's preparatory work for the UN conference was already having a visible effect on Latin American thinking, and perhaps anticipation of the larger conference explains in part why the results of the meeting stand as such an anticlimax to the Principles of Mexico. The Resolution of Ciudad Trujillo that the conference produced on

March 28, 1956 contained only one significant agreement: a legal definition of the continental shelf.

> The seabed and subsoil of the continental shelf, continental and insular terrace, or other submarine areas, adjacent to the coastal state, outside the area of the territorial sea, and to a depth of 200 meters or, beyond that limit, to where the depth of the superjacent waters admits of the exploitation of the natural resources of the seabed and subsoil, appertain exclusively to that state and are subject to its jurisdiction and control.[19]

This definition, although innovative in its combination of the two criteria, was based on earlier work of the ILC, and does not appear to relate to earlier work in the OAS context. On other key matters like the epicontinental sea, living resources of the seabed, the nature and scope of the special interest of the coastal state, and the breadth of the territorial sea, the conference could not reach agreement and therefore did not express an opinion.

Although the restraining influence of the United States on the proceedings should not be underestimated, the results of this conference indicate clearly that the Latin American countries still stood a considerable distance from a unified regional position on the law of the sea. While large majorities could be obtained for certain negative positions (like that against the three-mile territorial sea limit), and for vague formulations of extended coastal state rights (like those in the Principles of Mexico), it was not possible to get agreement on specific limits or specific definitions of rights. The agreement on the continental shelf, though a notable accomplishment, was made possible only by the nonspecific nature of the "exploitation" criterion.[20] All the Latin American states could easily reconcile this definition with their own existing continental shelf claims.

The Latin Americans' failure to achieve regional unanimity on the law of the sea does not detract from the fact that they made more progress toward this goal than any other group of states. The difficulties they encountered in trying to reconcile their differences demonstrated, on a small scale, the problems facing the world community. The whims of geography created a very different set of problems for the narrow-shelf countries on the west coast of the continent than they did for the wide-shelf countries on the east coast. These differences were enhanced by the mass of dissimilar unilateral legislation enacted by the various Latin American states. While this legislation was useful as a fund of models for possible regional action, it tended to work against regional thinking by entrenching national positions in concrete forms. Partly because of such factors, the

OAS's regional initiative was unable to gather more than a very general consensus in favor of increased coastal state control before it was overtaken and superceded by the wider effort of the United Nations. Neither the tenacity with which many of the states in this smaller forum defended their widely disparate positions, nor the pronounced move toward broad coastal state control over the seabed and the superjacent waters, boded well for the UN's prospects at the international level.

The International Response: The International Law Commission

The task of the ILC was on a larger scale than that of the OAS in that it covered both a wider range of issues and a larger number of states. While the IACJ had tried merely to reconcile the coastal state interests of OAS members, the ILC was mandated to codify the entire international law of the sea. The ILC itself was composed of 15 members, each nominated by his government and subject to approval by the UN General Assembly.[21] The commission awarded high priority to the law of the sea partly on the basis of the rapid development of national legislation, and it began its work in 1949 by choosing a rapporteur to report on the subject. At its meeting in 1950 the commission studied various aspects of the high seas and the continental shelf, and, on the request of the General Assembly, added the territorial sea to its agenda. It kept itself open to outside suggestions, and in this context received reports from, among others, the International Law Association and the International Bar Association.[22]

In 1950 the problem of defining the continental shelf was already beginning to exercise the commission. Since the subject was being considered under the heading of "high seas", there was some question as to whether the shelf beyond the territorial sea came under national or international jurisdiction. This dilemma was resolved fairly quickly in favor of the coastal states, but only on condition that coastal state powers be limited to control and jurisdiction over resources, and not involve sovereignty over the area itself. That decided, it remained to define some limit to coastal state jurisdiction. The 200-meter limit had not only the support of the various international law associations but also the powerful precedent of the Truman Proclamation to recommend it.

Another idea had been raised, however, by the American member of the commission. This was a suggestion that the limit be a flexible one based on the capability to exploit. The merits of this definition were that it did not conflict with existing national legislation, and that it would automatically keep pace with technology, thereby assuring an orderly regime of seabed exploitation, and avoiding the problem of a fixed limit becoming rapidly outdated. It also neatly sidestepped the problem of what regime would apply to the seabed outside a fixed limit, but it raised the specter of unlimited coastal state claims to the seabed following the march of technology. The main debate on this did not take place until the following year, by which time the commission had the added advantage of its rapporteur's report.

At the ILC's 1951 session the rapporteur, Mr. J. P. A. Francois, recommended the 200-meter isobath as the limit for the continental shelf. The commission initially accepted his definition, and even adopted it as the first article of a draft convention on the continental shelf. Further discussion, however, emphasized the instability of a fixed-limit definition, and eventually a subcommittee was appointed to reconsider the matter. This subcommittee concluded that a geographical limit was unsuitable because of its likely instability in the face of technological developments, and proposed the purely legal definition of exploitability as more appropriate. The commission accepted this argument, rejected an international agency for the continental shelf as too inefficient economically and too difficult politically, and made formal its rejection of coastal state sovereignty over the continental shelf. The commission also considered a combined depth/distance formula for determining the continental shelf limit. This was intended as a means of compensating those states with very narrow shelves, but was meaningless if restricted to the seabed because of the impossibility of exploiting deeper areas, and was far too contentious if it was to apply to fisheries because it raised the whole issue of a link between the continental shelf and the epicontinental sea that had caused so much controversy in Latin American legislation. The exploitability criterion proved a convenient compromise between those supporting the 200-meter limit (mostly the traditional three-mile limit states) and those wanting inclusion of a distance criterion (led by the Latin Americans).[23]

The result of the ILC's 1951 session was a set of draft articles on the continental shelf. Article 1 defined the continental shelf as "the seabed and subsoil of the submarine areas contiguous to the coast, but outside the area of territorial waters, where the depth of the superjacent waters admits of the exploitation of the natural resources of the seabed and subsoil." Article 2 restricted coastal state rights to "the exercise . . . of control and jurisdiction for the purpose of exploring it and exploiting its natural resources."[24] In its commentary on Article 1 the commission noted that "technological developments in the near future might make it possible to exploit resources of the seabed at a depth of over 200 meters," but it did not address the problem that its definition begged the question of a limit for the territorial sea. Neither did it define what exploitation meant: whether it applied only to individual situations, or whether it meant that the leading example set the standard for all; and what level of activity would constitute exploitation? The process of argument and compromise in the work of the ILC, and the lack of clarifying detail in its proposals, illustrate clearly the considerable extent to which the commission was a political forum.

Following its usual practice, the ILC requested comments on its draft from interested states, but only 18 replied, seven of which had members on the commission. Of the 18 states, only four supported the exploitation definition, while seven preferred 200 meters or some other fixed depth, and two wanted a distance criterion. It is noteworthy that both the United States and Chile, with their very different approaches to the subject, supported the exploitation criterion, and that Israel based her 1952 claim to the continental shelf on it. The comments on the nature of coastal state rights were almost evenly divided, with seven states favoring sovereign rights, six favoring "control and jurisdiction" rights, and the Netherlands alone supporting international control over exploitation of the continental shelf.[25]

At its 1952 session, the ILC concentrated on territorial sea and high seas questions; it did not return to the continental shelf issue until 1953. By that time it had had sufficient opportunity to absorb the comments made on its 1951 draft and to take note of new writing on the subject.[26] As a result of these other opinions the commission decided to reconsider the definition of the continental shelf. By a vote of seven to four with two abstentions, it abandoned the exploitation definition and returned to the 200-meter depth line. In addition, it made a slight response to the national comments by changing Article 2 to give the coastal state "sovereign rights for the purpose of exploring and exploiting" the natural resources (including sedentary species) of the continental shelf.[27] In its commentary on the articles, the commission defended the fixed-depth criterion by arguing its certainty of definition, its adequacy for current needs, and its parallel to the geographical realities of the continental shelf. Where those geographical realities created difficulties, as in the case of the deep trough off Norway's coast, the commission envisaged exceptions to the strict interpretation of the 200-meter rule.

For the next two years the ILC devoted itself to drafting articles on the territorial sea and the high seas. Except for an unsuccessful effort by the United States, Britain, and a few other states to have the General Assembly consider draft articles on the continental shelf separately from other law of the sea matters, the continental shelf issue lay dormant.[28] The commission was unable to make much headway in defining a uniform breadth for the territorial sea, though it did manage to restrict the debate to a three- to 12-mile range. This restriction, combined with the progress made on a definition for the continental shelf, reduced the importance of the territorial sea as a means of extending coastal state control over the seabed. Even a 12-mile limit was, almost everywhere, closer to the coast than the 200-meter isobath. Nevertheless, the territorial sea remained an important factor in seabed politics because of the possibility that inability to reconcile the deep differences on the question of breadth would wreck any attempt to achieve a package deal on the law of the sea. As if to underline this problem, states' comments on the ILC's draft articles on the territorial sea reflected an inflexible adherence to unilaterally established national limits.[29]

The commission's work on the high seas provided some clarification on seabed questions. After listing the traditional freedoms of the sea, the commission commented that "the list of freedoms of the high seas contained in this article is not restrictive. . . . There are other freedoms, such as freedom to explore or exploit the subsoil of the high seas and freedom to engage in scientific research therein."[30] This comment somewhat contradicted the continental shelf drafts since the shelf lay beneath part of the high seas, but it did give firm indication that the commission did not envisage a limitless extension of coastal state jurisdiction over the seabed whether under the exploitation criterion or by revision of fixed-limit definitions.

At its 1956 meeting the ILC presented final versions of all its draft articles on the law of the sea, and recommended that the General Assembly convene a conference to consider conventions on the matter. During this session the commission again reconsidered the definition of continental shelf limits, and again altered its previous draft. The initiative in this case came from the Latin American delegates, who had just returned from the OAS conference in Ciudad Trujillo. The definition of the continental shelf limits that they had achieved at Ciudad Trujillo was a strengthened version of the ILC's 1951 and 1953 draft articles combined, and they offered this combination for the commission's consideration. After lengthy debate, and the deletion of references to the continental terrace, the commission decided by a vote of nine to three with three abstentions to add the exploitation criterion to its 1953 draft.[31] The resulting draft Article 67 defined the continental shelf as

> the seabed and subsoil of the submarine areas adjacent to the coast but outside the area of the territorial sea, to a depth of 200 meters (approximately 100 fathoms), or, beyond that limit, to where the depth of the superjacent

waters admits of the exploitation of the natural resources of the said areas.[32]

In its draft Article 68 the commission retained unchanged the idea of sovereign rights for the purpose of exploring and exploiting the resources of the continental shelf, and continued to reject both the idea of coastal state soveriegnty over the shelf and the idea that the shelf should be internationalized. In its draft Article 69 the commission made explicit its view that "the rights of the coastal State over the continental shelf do not affect the legal status of the superjacent waters as high seas, or the airspace above those waters."

The many ambiguities in the combination definition of Article 67 clearly label it as a compromise solution. The commission admitted its lack of unanimity on this article, but the vote on it showed that it had more support than the simple 200-meter definition in the 1953 draft. The commission's commentary on Article 67 appeared to reflect a considerable polarization between the fixed and flexible limit schools of thought. Those in favor of a fixed limit appeared to think that technology would not overtake 200 meters for a long time, and that the advantages of a defined boundary therefore outweighed the problem of obsolescence. Those in favor of a flexible limit appeared to have a variety of motives. The Latin Americans had a group interest in promoting a definition that not only paralleled their regional position, but also did not conflict with their far-reaching national legislation. Some states perhaps supported the purely technical argument that the flexible limit would avoid the problem of obsolescence; and some might have supported the flexible limit as a way of avoiding fixed lines, because of a concern that such lines might become the framework for creeping jurisdiction into the water column.[33] The most obvious incentive for supporting the flexible limit, was, however, simply that it left open an opportunity for continued expansion of national jurisdiction on the seabed.

The commission's attempts to clarify Article 67 did not explain what significance the 200-meter criterion was supposed to have in the light of the exploitation criterion, and generally did not much reduce the vagueness of the proposal. No precise meaning was given to "adjacent" or "exploitation", and the relation of the legal concept of the continental shelf to the geographical concept was defined with sufficient ambiguity to sustain widely different interpretations of the absolute limit of coastal state jurisdiction on the seabed. The commission appeared to be trying to link its legal definition to the geographical concepts of adjacency and geological relatedness, but in its anxiety to cover fairly the geographical variety of national circumstances, it allowed consideration of interpretations as extreme as a division of the entire ocean floor among the coastal states.[34] It stated that it

> in no way holds that the existence of a continental shelf, in the geographical sense as generally understood, is essential for the exercise of the rights of the coastal State as defined in these articles. . . . Exploitation of a submarine area at a depth exceeding 200 meters is not contrary to the present rules, merely because the area is not a continental shelf in the geological sense.[35]

While this qualification appears to have been aimed at special circumstances within the bounds of the continental margin, it was itself sufficiently poorly qualified to admit of broader interpretation. The argument that the commission intended to limit the expansion of coastal state jurisdiction to the margin is supported by the commentary on Article 27 (freedom of the high seas). While still not resolving the paradox that the continental shelf itself underlay the high seas, the commission considered "continental shelf" and "exploitation of the subsoil of the high seas" as separate categories, and decided not to regulate the latter because it "had not yet assumed sufficient practical importance."

In its commentary on Article 68 (coastal state rights to the continental shelf), the commission made the important qualification that "the rights of the coastal State are exclusive in the sense that, if it does not exploit the continental shelf, it is only with its consent that anyone else may do so." This comment reflected a key element in states' interest in the continental shelf: namely, that in most cases the coastal state's interest was rather to deny freedom of access to others than to secure its own right of access.[36] At this time only the United States, Britain, and one or two other countries had any ability to undertake seabed mineral exploitation. The interest of all other coastal states was therefore to ensure that these technologically capable states could not freely mine the continental shelf regardless of the adjacent coastal state. This interest was well served by the ILC's draft, which gave the coastal state the necessary authority to control, and to extract benefits from, any seabed exploitation off its shores.

The ILC's draft articles on the law of the sea were important both as a coherent set of proposals on which the UN conference could base its deliberations, and as a statement of the prevailing international consensus on law of the sea issues. Like the Latin American countries, the ILC found it much easier to achieve a compromise on seabed limits than on limits in the water column. This is largely explained by the low level of economic interest in the seabed. With the exception of a few favored areas like the Persian Gulf and the continental shelf off California, most states did not, at this time, anticipate any significant income from seabed mining. In addition there were strong common sense and geographical arguments for giving the coastal state control over its shelf, a weight of unopposed precedent in support of coastal states exercising such control, and no weight of tradition relating to the seabed that went counter to the trend.

In the water column, by contrast, there were active economic interests almost everywhere, many of them in conflict; a vast weight of tradition relating to fishing, shipping, and military interests that militated against coastal state control; and a body of state practice that was heavily in dispute. The commission's drafts reflected this condition clearly. It was unable to narrow its proposals on the territorial sea beyond a three- to 12-mile window, and even within that range could find almost no agreement. It could not produce any consensus on the question of coastal state rights over fishing beyond the territorial sea, though in relation to this it was explicit in rejecting the link between the continental shelf and the superjacent waters that the Latin American states had been trying to establish in their unilateral legislation. Although the Latin American states had won their point on Article 67 (definition of the shelf) they had lost on both Article 68, which had refuted their claim to sovereignty over the shelf, and Article 69, which severed shelf rights from rights in the superjacent waters.

The ILC's proposals were a major landmark in seabed politics. They demonstrated that the type of claim pioneered by the United States in 1945 was a widely acceptable extension of coastal state jurisdiction. In purely economic terms, there was no dispute about the rights of the coastal state over the mineral resources in the seabed off its coasts. But while the ILC vindicated continental shelf claims that were motivated by an interest in seabed resources, it undermined such claims that were motivated either by a desire to extend sovereignty for its own sake, or by a desire to extend sovereignty in order to mount claims to the epicontinental sea. By tending to separate, in this way, seabed issues from other ocean issues, the ILC dampened much of the controversy that had surrounded continental shelf claims since 1945. This dampening effect was enhanced by the flexible definition of the shelf, which avoided any conflict over limits with existing or likely national claims. If the UN conference were to accept the ILC's proposals on the shelf, it looked as if the seabed would become an area of political, economic, and legal calm, above which the dispute over limits in the seas themselves would continue.

NOTES

1. The discussion in this section is heavily indebted to F. V. Garcia Amador, The Exploitation and Conservation of the Resources of the Sea (Leyden: Sythoff, 1963), pp. 2-28; Erin B. Jones, Law of the Sea: Oceanic Resources (Dallas: Southern Methodist University Press, 1972), pp. 6-77; Gerard J. Mangone, The United Nations, International Law, and the Bed of the Seas (Newark: University of Delaware Press, 1972), pp. 4-8; Brenda Horsfield and Peter Bennet Stone, The Great Ocean Business (London: Hodder and Stoughton, 1972), pp. 12-55; Sir C. J. B. Hurst, "Whose is the Bed of the Sea?", British Yearbook of International Law 4 (1923-24), pp. 34-43; and Ernest Katin, The Legal Status of the Continental Shelf as Determined by the Conventions Adopted at the 1958 United Nations Conference on the Law of the Sea, Ph.D. thesis, University of Minnesota, 1962, pp. 13-23. These sources provide a much fuller discussion of this period than is given here.

2. Edward Miles, "Technology, Ocean Management, and the Law of the Sea: Some Current History," Denver Law Journal 46, no. 2 (1969), p. 245.

3. For a more detailed background on the Truman Proclamation and full copies of the proclamation and the press release, see Marjorie M. Whiteman, Digest of International Law 4, U.S. Department of State Publication 7825, Washington, 1965, pp. 752-58; and Mangone, op. cit., pp. 8-9.

4. Whiteman, op. cit., p. 754.

5. Ibid., pp. 794-96. For background information on the many other continental shelf claims made between 1945 and 1950, see also pp. 789-813, and Richard Young, "The Continental Shelf in the Practice of American States," Inter-American Juridical Yearbook 1950-51 (Washington: Pan American Union, 1953), pp. 27-36. The origins of the 200-mile figure are obscure. Its relationship to fishery resources off the Chilean coast was far from precise, and no direct precedent existed. Indirect precedents existed in the Latin American context for considering coastal state jurisdiction well beyond the three-mile territorial sea, notably the Declaration of Panama (October 1939), which declared a 300-mile security zone around the 21 American republics, and a Uruguayan proposal (July 1940) for a 25-mile territorial sea for the whole continent. See Inter-American Juridical Committee, OEA/Ser.I/VI.2 CIJ-80 (Washington, D.C.: Pan American Union, 1966), pp. 24-30. Gerard Mangone (op. cit., p. 11) suggests that the figure originated in the U.S. neutrality legislation of September 1939, which mentions a 200-mile security zone. Whatever its origins (some have flippantly suggested it arose from a mistyping of 200 meters), the figure was

destined to be of seminal importance in the development of the law of the sea.

6. "The Abu Dhabi Oil Arbitration," in C. B. Bourne and L. G. Jahnke, Cases and Materials on Public International Law (Vancouver: University of British Columbia, 1972), pp. 7-43 to 7-47; and Myres S. McDougal and W. T. Burke, The Public Order of the Oceans (New Haven: Yale University Press, 1962), pp. 639-40.

7. H. S. Amerasinghe, "The Third World and the Seabed," in E. M. Borgese, ed., Pacem in Maribus (New York: Dodd, Mead, 1972), p. 239; Barry B. L. Auguste, The Continental Shelf (Geneva: Librairie E. Droz, 1960), pp. 238-41; E. Letts, "Peru," in L. M. Alexander (ed.), The Law of the Sea: The United Nations and Ocean Management (Kingston: University of Rhode Island Press, 1971), pp. 340-43; Alvaro de Stot, "The Latin American View of the Law of the Sea," India Quarterly 29, no. 2 (April-June 1973), p. 127; and F. V. Garcia Amador, Latin America and the Law of the Sea, Law of the Sea Institute Occasional Paper no. 14 (Kingston: University of Rhode Island Press, 1972), p. 33.

8. FAO Report on the Economic Implications of Sea Fisheries, UN Doc. A/CONF.13/16 (1957).

9. On the development of the argument linking the shelf with fishery resources, and relating both to the needs and responsibilities of the coastal state, see Auguste, op. cit., pp. 238-41; and Garcia Amador, op. cit., pp. 75-78. For a refutation of the "bioma" argument, see W. S. Wooster, "Scientific Aspects of Maritime Sovereignty Claims," Ocean Development and International Law Journal 1, no. 1 (Spring 1973), pp. 13-20.

10. It should be noted here that the United States was an important member of the OAS, and that the organization therefore cannot strictly speaking be considered a Latin American regional body. On the subject at hand, however, the Americans frequently found themselves in a small minority.

11. A copy of the declaration is reproduced in Shigeru Oda, The International Law of the Ocean Development (Leiden: Sijthoff, 1972), pp. 345-46.

12. This was 26,000 versus 1,234,700 square nautical miles, using data from U.S. Department of State, "Theoretical Allocations of Seabed to Coastal States Based on Certain UN Seabed Committee Proposals," International Boundary Study, Series A, Limits in the Seas, no. 46, Washington. A measure of the success the three governments had in defending their position was the enactment of the Reimbursement of Fines Act by the U.S. Congress in 1954 (68 Stat. 883, 83rd Congress, second session). This act provided for the repayment of fines imposed by Chile, Ecuador, and Peru on U.S. fishermen arrested in their zones. Since the United States

did not recognize the zones, it could not stop its fishermen from going into them, and did not choose to contest the issue militarily.

13. Andres Aguilar, "The Patrimonial Sea," in L. M. Alexander (ed.), The Law of the Sea: Needs and Interests of Developing Countries (Kingston: University of Rhode Island Press, 1973), p. 165; and the New York Times, August 24-26, 1955.

14. B. H. Oxman, "The Preparation of Article 1 of the Convention on the Continental Shelf," Journal of Maritime Law and Commerce 3 (1971-72), p. 445.

15. The five articles are reproduced in full in Inter-American Juridical Committee, op. cit., p. 33.

16. Ibid., p. 34.

17. A full copy of the principles is reprinted in the Inter-American Juridical Yearbook 1955-57 (Washington: Pan American Union, 1958), pp. 226-28, along with statements and reservations by some of the participating states, pp. 234-40.

18. The New York Times, April 1, 1956, IV, p. 8.

19. A full copy of the resolution is reprinted in Oda, op. cit., pp. 346-47; and Inter-American Juridical Yearbook 1955-57, op. cit., p. 261.

20. It is interesting to note that the New York Times on March 29, 1956 interpreted this shelf definition as being set by the leading state example—that is, the state which had exploited at the greatest depth from its shores would set the depth applicable to all other parties to the resolution. On this basis, it allocated a maximum 160-mile shelf to the United States because of American oil exploitation and the depth of the banks off the California coast.

21. During the period 1949-56, the members were from the following countries: the United States, Britain, France, the Soviet Union, Netherlands, Sweden, Greece, Czechoslovakia, Taiwan, Syria, India, Brazil, and Mexico. Panama and Colombia served 1949-53, and were replaced by Cuba and Bolivia in 1954.

22. The ILA recommended a basic 200-meter depth-line definition for the shelf with exceptions in cases where unusual geographic configurations created difficulty—for example, Chile. The IBA favored a 200-meter or geographical definition.

23. Oxman, op. cit., pp. 275-77, 281, 304. On the ILC's work generally, see McDougal and Burke, op. cit.

24. ILC Report for 1951, A/1858, pp. 17-20.

25. Ibid., for 1953, A/2456, pp. 42-72.

26. For example, M. W. Mouton, The Continental Shelf (The Hague: Martinus Nijhoff, 1952), p. 43. Mouton argued that 200 meters was a generous limit, unlikely to be exceeded by technology, while the exploitation criterion would be destabilizing because of the uncertainty it would create.

27. ILC Report for 1953, pp. 12-14; and Yearbook of the ILC for 1953, pp. 76, 83, 342-43.

28. Mangone, op. cit., p. 14.

29. ILC Report for 1955, pp. 25-49; and UN Doc. A/CN.4/99 and Adds. 1-9. The three- , four- , six- , 12- , and 200-mile positions were all represented in the comments.

30. Ibid., p. 3.

31. Yearbook of the ILC for 1956, p. 140.

32. ILC Report for 1956, p. 41.

33. This latter case is argued by Oxman, op. cit., p. 717.

34. See, for example, Shigeru Oda, "Proposals for Revising the Convention on the Continental Shelf," Columbia Journal of Transnational Law 7, no. 1 (1968), pp. 1-31.

35. ILC Report for 1956, p. 41.

36. This argument is discussed in Oxman, op. cit., p. 464.

CHAPTER

2

THE FIRST AND SECOND UNITED NATIONS CONFERENCES ON THE LAW OF THE SEA, 1958-60

NATIONAL CLAIMS UP TO 1960

The First and Second UN Conferences on the Law of the Sea grew out of a very different background from that of the 1930 League of Nations Conference. The process of national claims to jurisdiction beyond three miles that had begun after 1930 had, by 1958, reached quite considerable proportions. In the three decades between 1930 and 1960 36 states had made territorial sea claims beyond three miles, and 48 states had made some sort of claim to the continental shelf. Allowing for overlaps between these two categories, a total of 66 states had claims of some sort beyond three miles by 1960. Even allowing for the fact that 15 of the continental shelf claims had been made by Britain on behalf of colonies and dependencies, these figures meant that at least half of the states attending the UN conferences had a vested interest in obtaining recognition for extensions of coastal state jurisdiction beyond three miles. (See Table 2.1.)

The nature of the claims themselves indicated most of the problems that the conferences had to face. As of 1958, territorial sea claims were nearly evenly divided between those for and those against three miles. Some 38 states held to the traditional three-mile limit, while 34 others favored four or more miles.[1] In the latter group 12 states claimed six miles and 14 claimed 12, but while El Salvador was the only state claiming a 200-mile territorial sea, at least four other Latin American countries claimed 200-mile zones that were close to being territorial seas. The question of the breadth of the territorial sea was thus not only a very open and contentious issue for the conferences; it was also still an important aspect of national jurisdiction over the seabed.

TABLE 2.1

Territorial Sea Claims Beyond Three Miles of 36 States Up to 1960

	4 Miles	5 Miles	6 Miles	9-10 Miles	12 Miles	200 Miles
Western states	Finland Sweden[a] Norway[a] Iceland[a, b, c]		Greece[a, c] Italy[a] Israel[c] Spain[c] Portugal[c]			
Soviet group			Yugoslavia	Albania	Bulgaria Romania USSR[a] China	
Latin American states			Uruguay[a] Colombia[a] Cuba[a]	Mexico[a, c]	Guatemala[a, c] Venezuela[a, c] Ecuador[b, c] Panama[b, c]	El Salvador
Arab states					Libya[c] Iraq[c] Saudi Arabia[c] Egypt[c] Sudan	
Asian states		Cambodia[c]	India[b, c] Ceylon[b, c] Thailand		Indonesia Iran[a, c]	
African states					Ethiopia	

[a]Claim made before 1945
[b]State with fisheries claim beyond 12 miles
[c]State with both continental shelf and territorial sea claim
These symbols apply also to Table 2.2

Source: Data derived from a wide variety of sources, the most comprehensive of which is U.S. Department of State, Office of the Geographer, Bureau of Intelligence and Research, Limits in the Seas No. 36: National Claims to Maritime Jurisdiction, April 1974.

Continental shelf claims up to 1958 were also very uneven in nature. The great majority (33) were nonspecific in terms of their definition of the continental shelf, relying usually on an unadorned reference to the shelf or the terrace. Five used some version of the 200-meter/100-fathom/600-foot formula, and a sixth chose the 50-meter depth line. Five used the exploitation criterion devised by the ILC (India's claim dates from 1959), and a maximum of six, if Honduras and El Salvador are counted, claimed 200 miles. These claims were also mixed in terms of degree, ranging from full sovereignty over the shelf to jurisdiction over the resources, but in terms of seabed resources only, none of them left any doubt that the coastal state was to have exclusive rights of exploitation.

The question of coastal state rights over fishing was related to both the continental shelf and territorial sea issues. The Latin American and Icelandic argument linking the continental shelf and the epicontinental sea was still far from being a dead letter, and at least a dozen states had made claims to fisheries beyond 12 miles. Most of these were between 100 and 200 miles in extent, and so represented a major assault on the accepted traditions.

On the territorial sea and fishing issues, national claims were the most obvious basis on which states could align in the conference, and the work of the ILC had not provided any alternative compromise. On the continental shelf, however, the ILC had provided a compelling definition that had great potential as a compromise because of its flexibility. Thus, except for the question of the nature of rights within the limits of the shelf, national positions on the continental shelf were less important going into the conference than were national positions on the territorial sea and fisheries.

In addition to alignments based on national claims, the UN conference was also subject to other alignments. The conference was held during the height of the cold war, and so experienced the strong pressures of Soviet-Western polarization. The Soviet group generally followed the Soviet Union in supporting a 12-mile territorial sea, and along with the group's anti-colonial stand, this created considerable common ground between the East European states and the substantial number of Arab, Asian, and Latin American states that also favored 12 miles. The Western group, including Western Europe, the white Commonwealth countries, and North and South America, was deeply divided on many law of the sea issues. It contained most of the traditionalist three-mile states, but could hardly be said on this issue to include the Latin Americans, who definitely opposed continuation of a regime based on narrow coastal state jurisdiction. Thus cold war and law of the sea alignments reinforced each other only inasmuch as the principals in the former dispute were on opposite sides in the central conflict over the breadth

TABLE 2.2

Continental Shelf Claims of 49 States, 1945-60

	Unspecified Geological	50 Meters	200 Meters or Variants	Exploitation (with or without 200 Meters)	200 Miles
Western states	Iceland[b, c]		United States	Israel[c]	
	Australia		Portugal[c]		
	Spain[c]				
	Greece[c]				
Soviet group					
Latin American states	Argentina[b]		Mexico[c]	Venezuela[c]	Chile[b]
	Panama[b, c]		Brazil	Honduras[b]	Peru[b]
	Guatemala[c]				Costa Rica[b]
	Nicaragua				Honduras ?[b]
	Dominican Republic				Ecuador[b, c]
	Guyana				
	Trinidad				
	Jamaica				
	Bahamas				
	British Honduras				
	Falkland Islands				
Arab states	Nine Persian Gulf sheikdoms			Egypt[c]	
	Saudi Arabia[c]			Morocco	
	Libya[c]				
	Iraq[c]				
	Kuwait				
Asian states	Philippines	Cambodia[c]	Pakistan	India[b, c]	
	Iran[c]				
	South Korea[b]				
	Ceylon[b, c]				
	Burma				

[b, c]See Table 2.1.

Source: Data derived from a wide variety of sources, the most comprehensive of which is U.S. Department of State, Office of the Geographer, Bureau of Intelligence and Research, Limits in the Seas No. 36: National Claims to Maritime Jurisdiction, April 1974.

of the territorial sea. Following the custom of the time, both cold war blocs sought support for their positions among the less committed states of the third world, and on this issue the pattern of historical claims made that search more fruitful for the Soviets, and more problematical for the Western states, than was the norm.

One other aspect of national claims that might have been, but was not, an element at the conference was the problem created by disputes between states over claims. A few such disputes had occurred, almost all of which were between distant water fishing states and states claiming to protect their coastal fish stocks, but no major disputes had arisen over delimitation between states that had adjacent or opposite coasts. The Soviet Union signed territorial sea delimitation agreements with Norway (1957) and Poland (1958), and Saudi Arabia and Bahrein agreed on a continental shelf delimitation (1957), but aside from these three cases this aspect of extended national claims was not important.

DEVELOPMENTS IN MARINE TECHNOLOGY AND OCEAN ACTIVITY, 1956-60

National claims were not the only source of pressure on the UN Law of the Sea Conferences. Many of the claims themselves had been stimulated by the increasing scale and variety of human activity in the oceans, and these changes in the pattern of ocean use were an important factor underlying the deliberations. Increases in the scale of ocean use occurred most visibly in relation to fishing and shipping, and were largely the result of steady growth in the size and wealth of populations and in the number of independent states. The tonnage of world sea trade experienced an annual increase of 8 percent from 1950 to 1958, and the weight of the world fish catch burgeoned from 21.1 million metric tons in 1950 to 36.7 metric tons in 1959. The expansion of shipping resulted in increased concern over pollution by oil (although this was not anywhere near the level of a general environmental concern), and the dramatic rise in the fish catch created considerable alarm over the depletion and unfair sharing of coastal fish stocks. Effects such as these undermined the assumption of the inexhaustibility of ocean resources on which the regime of freedom of the seas was based, and, as noted, raised the incentives for coastal states to seek wider jurisdiction.

Increases in the variety of ocean activity were strongly related to technological developments that increased human capability to operate in the ocean environment. Technology of this type came from three sources: naval research and development, marine science research, and the offshore oil industry. Naval involvement in marine research at this stage was largely concerned with submarines, although World War II had also resulted in improved scuba and pressure-suit diving equipment. The U.S. Navy began using sonar devices located on the seabed in the early 1950s, and by the middle of the decade had built the first nuclear-powered submarine. The USN was also interested in deep-diving vehicles, and in 1958 purchased the bathysphere Trieste built by the French oceanographer Auguste Piccard.

Oceanographers were approaching the end of a long period in which their main activities were surveying, mapping, and data collection. As a result, the outlines of the tectonic plate theory were beginning to emerge, and oceanographic activities in general were becoming more visible through such events as the International Geophysical Year in 1958. Nevertheless, while these military and oceanographic activities were important in terms of the scientific and technological imperatives they were creating, they had not reached the stage at which they could become a major influence on seabed politics.

At this time, that role was reserved for the offshore oil industry. The industry had come a long way from its humble beginnings in the 1890s, and by 1958 was in a state of rapid development and expansion. Offshore drilling had been pushed out over a mile from shore in the decade before the war, but was restricted to depths of not much more than a dozen feet. The first well beyond the three-mile limit was not drilled until 1947, off the coast of Louisiana, and even then lay in only 17 feet of water. Deep water drilling technology did not begin to change the face of the industry until the 1950s, and when it did become available it spread quickly. A major innovation that enabled this spread to take place was the invention of a floating, mobile drilling platform, which was first successfully tested in 1947. The floating platform liberated the industry from clumsy bottom-fixed drilling rigs, and so made the move to deeper waters possible.

The year 1947 was also significant as the last year in which the United States was a net exporter of oil. This change in status added to the incentives for seeking oil on the continental shelf—especially since offshore oil technology was almost entirely in American hands at this time. By 1953 drilling was taking place in depths up to 70 feet and at distances up to 25 miles from shore. By 1958 drilling rigs were operating in depths approaching 300 feet, and the first experimental deep water drilling ship had extracted cores from the seabed at a depth of 1,500 feet. At least ten giant fields (more than 500 million barrels of recoverable oil) were discovered on the continental shelf up to 1958, and between 1954 and 1957 the number of oil installations in the Gulf of Mexico nearly trebled from 238 to

665. Despite the relatively high costs of deep water drilling, a report submitted to the UN conference in 1958 reported a consensus within the industry that commercial drilling would be possible in depths up to 600 feet within a decade.[2]

Up to 1958, the expansion of the offshore oil industry, though dramatic, was restricted to a small number of areas—principally Venezuela, California, the Gulf of Mexico, and the Persian Gulf. Although the industry's exploration had considerably raised expectations of wealth from the continental shelf, the geographical concentration of the finds had not encouraged the worldwide hopes for continental shelf oil that were to arise later. Despite this, the growth of the offshore oil industry after World War II contributed greatly to undermining the assumption of inexhaustibility of resources that supported the old freedom of the seas regime. Oil resources were quite plainly exhaustible, and were not subject to the same arguments that were applied, albeit with diminishing force, to fish.

The characteristic of exhaustibility went a long way toward explaining the relative readiness (compared to fishing and the territorial sea) with which extensions of coastal state jurisdiction over continental shelf resources found acceptance in the international community. Likewise, the rapid development of deep drilling capability explained the dilemma over fixed versus flexible limits in the ILC. Those who argued in favor of fixed limits could point to the enormous increase in costs and technical difficulties that would accompany a move by the oil industry beyond 200 meters depth. Those who favored flexible limits could argue with equal force that the previous decade demonstrated the facility with which the industry could overcome precisely that type of problem. In any event, both sides were arguing about offshore oil technology, and neither suspected that the big breakthrough would eventually come from the hard mineral industry.

The same argument of exhaustibility applied to hard minerals as to oil, but up to 1958 the offshore hard mineral industry had made little impact on seabed politics. While the industry enjoyed a steady growth, its technology remained relatively unchanged, and it experienced no massive expansion comparable to that in offshore oil. For the most part its operations were confined to areas close to the shore and to water depths of less than 100 feet. Bucket dredges were not practical in depths over 150 feet and were vulnerable to rough weather. Grab dredges, while capable of operating at almost any depth, became uneconomic at much over 200 feet because of the time involved in raising and lowering the grab. Suction dredges were still at a crude stage of development, and were restricted in depth by problems of pressure. Thus, although resources such as manganese and phosphorite nodules were known, lack of demand for them and limitations in the technology meant that offshore hard mineral mining was not among the early incentives for expanding coastal state jurisdiction over the seabed.

These developments in marine technology and ocean activity were largely responsible for the emergence of the coastal state as a major actor in ocean and seabed politics discussed above. This process was reinforced by the vigorous nationalism of the third world countries and by their own increasing needs and capabilities, and was farthest advanced in Latin America. It created conflicting interests between coastal and maritime states, and within countries like the United States which had strong interests of both types. Inevitably it also heightened the awareness of the landlocked states, which found themselves being left out of even more wealth and opportunity than had been the case before. Although the emergence of coastal state self-awareness was far from complete by 1958, the tide of national claims since 1945 indicated that it was sufficiently developed to ensure that the UN Conference on the Law of the Sea would not be a mere ritual codification of the traditional rights of the maritime states.

THE 1958 AND 1960 CONFERENCES

For the purposes of this study the two UN Law of the Sea Conferences can be considered as one event. Although they were separated by two years, the group of states participating in them underwent little change, and the agenda of the second conference was merely a continuation of unsettled leftovers from the agenda of the first. Eighty-six states participated in the 1958 session, and 84 of these, plus four additional states, made up the 88 attending in 1960. (Afghanistan and Nepal attended in 1958 but not in 1960, and Cameroon, Guinea, Ethiopia, and Sudan, three of which were newly independent, attended in 1960 but not in 1958.) The 1958 conference considered all aspects of the law of the sea and produced four conventions on the subject. The 1960 conference took up the questions of the breadth of the territorial sea and of coastal state jurisdiction over fisheries, neither of which had been settled in 1958. It was unable to find a consensus on either question and did not produce a convention.

The composition of the group of states at the 1958 conference is worthy of note. Twenty-nine were Western states, ten belonged to the Soviet group, 20 were Latin American, nine were Arab, 16 were Asian, and only two were African. In terms of developed and developing states this represented respective totals of 38 (the Western and Soviet groups minus Turkey) and 48. Cold war mentality and the dominance of the Western states in the United Nations were clearly reflected by the fact that China, East Germany, North Korea, and North Vietnam were not present while Taiwan, West Germany, South Korea, and South Vietnam were. Thirty-two states, well over a third of those present, had national claims that exceeded the prevailing norms (a territorial sea, continental shelf, fishing, or contiguous zone claim beyond 12 miles). Of these only six were developed states (all Western), while 26 were developing countries (14 Latin American, three Arab and nine Asian). In geographical terms relating to the seabed, 12 of the participants were landlocked, 12 shelflocked, and 60 open coast (excluding the Ukrainian and Byelorussian republics). Nearly 70 percent of the participants had continental margins exceeding 10,000 square nautical miles in area.

The Fourth Committee of the 1958 conference was allocated the continental shelf as its subject. This was a committee of the whole, and any proposal passed within it by a simple majority would be recommended to the plenary session of the conference, where it would have to obtain two-thirds approval of those voting yes or no. The Fourth Committee had before it the relevant draft articles recommended by the ILC, of which Article 67 on the definition of the shelf, and Articles 68 and 69 on the nature of states' rights over the shelf, were unquestionably the most important. (See Chapter 1.) States were, however, free to propose alternative drafts to those of the ILC, and more than a dozen did so.

On the question of the definition of the continental shelf, 15 alternatives or amendments to Article 67 were raised in the Fourth Committee. Many of these were withdrawn in favor of similar proposals, but with one exception all the major alternatives came to a vote. The exception was a memorandum by West Germany (L.1)[3] opposing the whole concept of coastal state rights on the continental shelf beyond the territorial sea. As a shelflocked state and a supporter of the traditional three-mile limit, West Germany had nothing to gain from the continental shelf doctrine. The fact that L.1 received no support (except from Monaco) indicated the wide extent to which state practice had created acceptance of the shelf doctrine; the issue was not whether, but how much and with what rights.

West Germany, and in their own way Chile, Ecuador, and Peru, were thus the only ones to raise the issue that the continental shelf doctrine was randomly discriminatory on the basis of geographical pot luck. Some states received huge areas and others received nothing. Some wealthy states, like the United States, received resource-rich areas, while some poor states, like Chile, received barren areas despite their long coastlines. The whole question of injustice inherent in the continental shelf doctrine was not, with these few exceptions, an important aspect of the continental shelf debate at the 1958 conference.

The proposals for legal limits to the continental shelf were more varied than those considered by the ILC, although the essential question remained whether the limit was to be fixed or flexible. On the side of fixed limits, only France, Italy, South Vietnam, and Lebanon supported a 200-meter line, and a French proposal to this effect (L.7) was roundly defeated 12 to 48 to 7. Since 200 meters was evidently considered inadequate as a fixed limit, the only option for those who favored such a limit was to devise definitions that would give more to the coastal state and thereby overcome the problem of early obsolescence. The Netherlands, Britain, India, Sweden, and Canada all spoke in favor of a 550-meter limit, but an Indian proposal (L.29 rev.1), although faring much better than the French suggestion, was rejected 21 to 31 to 16. Yugoslavia tried to develop a combined depth/distance formula, but its effort received no encouragement, and when brought to a vote was supported by only two states. Panama, Spain, Canada, Guatemala, Egypt, and the Philippines all supported the idea of matching the legal definition of the shelf to the geographical reality, but a Canadian proposal (L.30) based on a depth/gradiant break formula was defeated twice, using different depths, and never received more than 16 supporting votes.

Those in favor of a flexible limit supported either the ILC formula or the exploitation criterion. More than a dozen states, including the Soviet Union, India, and eight Latin American states, spoke in favor of a flexible limit, but France, Britain, Canada and several others spoke against the exploitation criterion. A South Korean proposal (L.11) to use the exploitation criterion alone was, somewhat surprisingly, rejected decisively by 13 to 42 to 13. Since none of the "pure" fixed or flexible limit proposals could attract a majority, the Fourth Committee was in need of a compromise, and the ILC's formula was the obvious choice. After the acceptance of a Philippine amendment (L.26), which stipulated that the definition should apply to islands as well as continents, the ILC's 200 meters plus exploitation formula was put to a vote and accepted by 51 to 9 to 10.[4] In comparison with the vote on the South Korean proposal, this vote indicated the importance that was attached to the 200-meter element in the ILC's formula as a limiting factor, despite the fact that the significance of 200 meters in the combined proposal was unclear. It seems evident from this that many of the states supporting the ILC formula as a compromise did so in the belief that 200 meters was in some way a significant limitation on the exploitability criterion.

The vote on the ILC's amended draft Article 67 was a compromise made as a result of the fact that most states favored the continental shelf doctrine and wanted to see it incorporated into international law. The lack of response to the West German attack on the doctrine supports this view, and the big majority in favor of the ILC formula indicated the extent to which states were prepared to compromise their preferences on limits in order to achieve a broad acceptance of the doctrine as a whole. The fact that no alignments on either a political or a geographical basis were visible also supports the interpretation of the vote as a compromise. The United States and the Soviet Union both voted for it, and many of those voting against it did so because of their opposition either to the doctrine (West Germany) or to the open-endedness of the exploitation criterion (France, Italy, Pakistan). South Korea voted against because of its opposition to the 200-meter criterion. Western and Asian states made up the great majority of those opposing or abstaining, but both groups also had equal or larger contingents supporting. States with similarities in all the relevant attributes, including existing national legislation and area of geographical margin, were to be found on both sides of this vote.

A somewhat clearer view of the compromise represented by the vote in the Fourth Committee was obtained when the issue of exploitation was raised again in the plenary meeting of the conference. A separate vote was held on the exploitation clause, and although it was accepted by 48 to 20 to 2, the 20 opposing votes, when compared with the votes in the Fourth Committee, revealed which states had accepted the compromise despite their dislike of the exploitation criterion.[5] Eight of the 20 negative voters (Denmark, Finland, New Zealand, Norway, Portugal, Sweden, the Byelorussian republic, and Yugoslavia) had accepted the compromise in the Fourth Committee. Along with the others who voted against the exploitation clause, these revealed an almost solid bloc of 15 Western European states (plus Japan, New Zealand, Yugoslavia, the Byelorussian republic, and Pakistan) opposed to the flexible limit. This was in line with the extremely conservative view of extensions of coastal state jurisdiction generally held in Europe on the basis of the long tradition of dominant maritime status. The compromise was, however, reinstated by the plenary meeting when it passed Article 67 as recommended by the Fourth Committee by a vote of 51 to 5 to 10. On Article 67, then, the combination of a flexible definition that did not conflict with any existing legislation, and a widespread acceptance of the basic idea of the continental shelf doctrine, resulted in a broad consensus along the lines originally developed by the OAS and the ILC.

The debate on the nature of states' rights over the shelf—Articles 68 and 69—centered on two issues: first, whether or not states' rights should be described in terms of sovereignty or in terms of exclusive jurisdiction; and second, whether or not the definition of natural resources to which these rights would apply should include living resources or be restricted to mineral resources. Both of these questions were important, sovereignty because it reflected on many of the claims and protests already made concerning the continental shelf, and living resources because it involved the link between the shelf and fisheries that had already become controversial in the context of Latin American legislation.

On the question of sovereignty, the Fourth Committee was heavily divided between those favoring complete territorial sovereignty over the shelf for the coastal state and those strenuously opposing such sovereignty. This division went largely along developing/developed lines, with the Latin American states prominent among those favoring sovereignty and the European states prominent among those opposing it. A proposal by Mexico (L.2) favoring the sovereignty position was defeated 24 to 37 to 6, and this made it clear that a substantial majority of states were still not prepared to acquiesce in the more extreme claims of the Latin Americans. This vote effectively disposed of the extreme position, and thereby opened the way for the ILC's compromise wording, which involved sovereignty only over the exploration and exploitation of resources.

Even this degree of sovereignty, however, was offensive to the United States, which proposed that the term "sovereign rights" be replaced by the term "exclusive rights" (L.31). When this was put to a vote, it passed by the narrow margin of 21 to 20 to 27. The United States received support from the hard core of European traditionalist states and from a few of the developing countries with which it had close links (Philippines, Liberia, Taiwan). The Soviet group and a miscellany of other states opposed, but the vote was remarkable not for the alignments it revealed but for the very large proportion of abstentions. The Fourth Committee recommended to the plenary the ILC draft of Article 68, as amended by the American proposal, by a vote of 34 to 14 to 17.

Under the circumstances, it seemed doubtful whether the amended Article 68 would be able to attract a two-thirds majority, and before the matter came to a vote in plenary the United States representative agreed to withdraw the amendment in favor of the original ILC draft. This was done because the fear of extensions of continental shelf sovereignty of any sort into the superjacent waters, which was the motive behind the amendment, was largely allayed by the progress of negotiations on Article 69.[6] On this issue, then, as on the issue of limits, the basic lack of hostility to the continental shelf doctrine, combined with the existence of a workable ILC compromise, resulted in a widely accepted draft. The plenary passed Article 68 by a vote of 59 to 5 to 6.

On the question of defining the natural resources to which these rights would apply, the extreme positions were on the one hand to restrict rights to mineral resources only, and on the other to include not only minerals but also sedentary species and bottom fish. The middle position was to include minerals and sedentary species, but not bottom fish. This dispute related much more to fishery than to seabed politics, and it suffices here to note that proposals on the extreme positions were all defeated by large majorities, and that a compromise proposal representing the middle position (L.36) was adopted 41 to 11 to 17. One other successful amendment to Article 68 was proposed by Cuba (L.45); it provided that coastal state rights on the shelf did not depend on occupation or proclamation. This was important in terms of guaranteeing that states would exercise their rights regardless of whether it was they or others who undertook exploitation. The Cuban proposal was passed 41 to 7 to 12, and accepted along with L.36 in the version of Article 68 approved by the plenary.

The defeat of the territorial sovereignty position in Article 68 made more or less inevitable the adoption of Article 69. The ILC's version of this article had read, "The rights of the coastal state over the continental shelf do not affect the legal status of the superjacent waters as high seas or that of the air space above those waters." This article was a direct assault on those states, many of them Latin American, that had linked fishery claims to sovereign continental shelf rights. Argentina made an attempt to forestall this attack by proposing (L.6) that continental shelf rights be restricted only in regard to the freedoms of navigation and overflight, but this was defeated 10 to 40 to 10. The ILC's draft was passed by the Fourth Committee by 54 to 0 to 8 and in the plenary by 43 to 0 to 3.

On the continental shelf issue, then, the results of the 1958 conference were very much in line with the draft proposals evolved by the ILC in 1956. Although the conference had rebuffed the attempts of a few states to claim complete sovereignty over their continental shelves, it had strongly confirmed the continental shelf doctrine. In terms of seabed interests alone, the coastal states had been given not only sovereign rights to the resources, but also license to exploit them as far out as they were able to. In comparison with other issues at the conference, the continental shelf involved little contention. The basic idea was widely accepted, and the debate was confined almost exclusively to haggling over the details of implementation and interpretation. Both the United States and the Soviet Union firmly supported the final version of Articles 67, 68, and 69, and the leader of the American delegation pronounced that the Continental Shelf Convention was "highly satisfactory" to his country.[7] The convention as a whole was passed in the plenary by a vote of 57 to 3 to 8, with only West Germany, Belgium, and Japan opposing it.[8]

Given the state of seabed technology and interests at the time, it appeared as though the 1958 Convention on the Continental Shelf had set seabed politics more or less permanently to rest. Not only had a widely acceptable formula been found with which to operationalize the continental shelf doctrine, but the adoption of a flexible limit had largely obviated the question of a regime for the seabed area beyond the limits of national jurisdiction. National limits would simply expand as capability improved, and since interest in the outer continental shelf was restricted entirely to oil, there was no reason to expect that seabed exploitation would develop other than by a slow and steady move to the limits of the shelf and then down the slope to the rise. Nobody foresaw any interest in the deep seabed for a very long time to come, and so the ambiguities in the terms "adjacent" and "exploitation" that the 1958 conference inherited from the ILC remained unclarified. It was not clear whether the deep seabed came under the High Seas Convention, as the ILC had apparently intended (in which case deep seabed mining would come under the inter alia provision of the freedoms of the high seas) or whether it came under the Continental Shelf Convention (and therefore under coastal state control, as could be argued from an interpretation of the exploitation criterion and the ILC's commentary on it).

In either case, the 1958 conventions appeared likely to put an end to speculations about some form of international regime for the seabed. The idea of such a regime, with appropriate organizational machinery, has been traced back as far as 1934, when it was mentioned in a resolution of the Institut de Droit International,[9] but interest did not become general until the ILC began its work in 1950. In that year the Copenhagen Conference of the International Law Association considered it, and by 1951 it had cropped up in the ILC's debates. In 1953, the international regime option was considered in the report of the ILC's rapporteur on the high seas, and the Dutch government advocated an international organization for the continental shelf in its response to the ILC's early drafts on the subject. The ILC eventually rejected the idea of an international regime for the continental shelf, and felt that a regime for the deep seabed was not yet a matter of sufficient urgency to warrant its attention.

Despite this, the idea came up again at the 1958 conference. India raised it in connection with the discussion of fixed limits,[10] and West Germany raised it in the context of rejecting coastal state control over the shelf (L.1). More importantly, Monaco raised the possibility of an international office of the sea.[11] All of these proposals, however, were lost in the general enthusiasm for coastal state control over the continental shelf. Coastal state control was a relatively simple and uncontroversial solution to the problem and seemed to be the regime most likely to encourage exploitation activity—a development that was generally desired. Internationalization, though noble and equitable in concept, offered every prospect of being extremely difficult to put into operation. No preparatory work on the subject had been done, and very few states seemed willing to entertain a complicated solution where a simple one would suffice. Thus, even on this small front, the 1958 convention seemed to have settled the issues.

As mentioned before, the question of extending coastal state jurisdiction by broadening the territorial sea did not have the backing of common sense, unopposed precedent, or absence of countervailing tradition that characterized extension of coastal state jurisdiction over the continental shelf. For economic and historical reasons the breadth of the territorial sea was an extremely controversial issue, and this controversy went a long way toward explaining the sensitivity on state sovereignty over the continental shelf that was manifested in the Fourth Committee. It likewise explained the specific denial of any link between the seabed and the superjacent waters contained in Article 69 (Article 3 of the final version of the convention), and the necessity for states to accept this separation if any progress was to be made toward a general agreement on the continental shelf.

The success of the Fourth Committee's work largely removed the question of the territorial sea from the realm of seabed politics, but one important link between the territorial sea and seabed limits did remain: the fact that the inner boundary of the legal continental shelf was to be measured from the outer edge of the territorial sea. Since virtually the entire debate on the breadth of the territorial sea took place within the framework of a limit somewhere between three and 12 miles, this loophole did not have much practical significance. But since neither of the UN conferences was able to set a fixed limit for the territorial sea, a possibility remained that coastal states could make claims that would include the seabed beyond the limit set in the Continental Shelf Convention. The existence of El Salvador's 200-mile territorial sea claim, and the 200-mile zones of Chile, Ecuador, Peru, and Costa Rica exemplified this possibility.

The territorial sea question at the 1958 and 1960 conferences is also of relevance because it revealed, much more clearly than did the debate in the Fourth Committee, the nature of alignments on ocean politics as a whole. Territorial sea and coastal fisheries limits were dealt with together as a package that combined the most crucial aspects of coastal state jurisdiction. The package was necessary because to a large extent it was states' concern over fishing that motivated wider territorial sea claims. At the 1958 conference, proposals were made in favor of territorial sea maximum widths of three miles, six miles, six miles plus an additional six mile fishing zone, 12 miles, 12 miles plus an additional contiguous zone, and the right of the coastal state to determine its territorial sea width unilaterally. All the three-mile proposals were withdrawn when it became obvious that they had no chance of attracting a majority. The only six-mile proposal to come to a vote was defeated 16 to 49 to 4. A six-plus-six proposal by the United States (L.29) received a majority of 45 to 33 to 7 in the plenary but this fell short of the two-thirds majority necessary for approval. A 12-mile proposal by eight developing countries (L.34) likewise received a majority in the plenary (39 to 38 to 8) but failed to get two-thirds by an even wider margin. Proposals for anything beyond 12 miles were all defeated, and the Peruvian delegation withdrew its unilateral proposal, which was based on the Principles of Mexico, because of an obvious lack of support.

At the 1960 conference, proposals were made only on the six-plus-six and 12-mile variations. A 12-mile proposal sponsored by 18 developing countries and supported by the Soviet Union (L.2 rev.1) was defeated 36 to 39 to 13 in the Committee. The six-plus-six proposal sponsored by the United States and Canada (L.11) received a majority of 54 to 28 to 5 in the plenary, thus failing by one vote to achieve the necessary two-thirds. Even had this proposal passed, it seems unlikely that the large group of 12-mile states led by the Soviet Union would have abandoned its position or that, in many cases, the individual states would have abandoned existing legislation. Because of the cold war division, the vote would have been interpreted as political and not representative of a true international consensus on the subject. In terms of actually preventing wider claims to the territorial sea, it would probably have represented a victory on paper only for the narrow-limit states—although such a victory would unquestionably have been useful as propaganda against those with 12-mile claims.[12]

Table 2.3* shows how states were aligned on this issue. To a considerable extent alignments followed existing national legislation,

---

*The data on this table for 12 miles were derived from states' votes on the eight-power proposal (L.34) in the plenary session of

TABLE 2.3

Alignment of States at the 1958 and 1960 LOSCs: Voting on 12 Mile and Six-Plus-Six Mile Proposals

	1	2	3	4	5	6
Western	Australia (3)	Finland (4)			Iceland (4)	
states	Austria-	Holy See-				
	Belgium (3)	Japan (3)				
	Canada (3)					
	Denmark (3)					
	France (3)					
	West Germany (3)					
	Greece (6)					
	Ireland (3)					
	Israel (6)					
	Italy (6)					
	Luxembourg-					
	Monaco (3)					
	Netherlands (3)					
	New Zealand (3)					
	Norway (4)					
	Portugal (6)					
	San Marino-					
	South Africa (3)					
	Spain (6)					
	Sweden (4)					
	Switzerland-					
	Turkey (?)					
	United Kingdom (3)					
	United States (3)					
Latin	Brazil (3)	El Salvador (200)	Argentina (3)		Chile (30?)	
American	Cuba (6)	Nicaragua (?)	Bolivia-		Ecuador (12)	
states	Dominican Republic (3)		Colombia (6)		Mexico (9)	
	Haiti (6)?		Costa Rica (?)		Panama (12)	
	Honduras (?)		Guatemala (12)		Peru (?)	
	Paraguay-		Uruguay (6)		Venezuela (12)	
Asian	Pakistan (3)	Laos-	Malaysia (3)	Afghanistan-	Burma (?)	Philippines (?)
states	South Vietnam (3?)	South Korea (?)	Ceylon (6)	India (6)	Indonesia (12)	
	Thailand (6)			Iran (12)		

	Taiwan (3)			Cambodia (5) Nepal-	
African states	Liberia (3)		Ethiopia (12) Ghana (?) Cameroon (?)		
Arab states			Jordan (3) Tunisia (3)		Egypt (12) Iraq (12) Lebanon (?) Libya (12) Morocco (?) Saudia Arabia (12) Sudan (12) Yemen (?)
East European states				Poland (3?)	Albania (10) Bulgaria (12) Byelorussia SSR- Czechoslovakia- Hungary- Romania (12) Ukraine SSR USSR (12) Yugoslavia (6)

1. Voted no on 12 miles and yes on six-plus-six. Therefore strong support for narrow limits and strong opposition to wide limits.
2. Voted no or abstained on 12 miles and respectively abstained or voted yes on six-plus-six. (Therefore a moderate position favoring narrow limits and opposing wide limits?)
3. Voted yes on both proposals. Therefore no strong feelings either way.
4. Voted yes or abstained on 12 miles and respectively abstained or voted no on six-plus-six. (Therefore a moderate position favoring wide limits and opposing narrow limits?)
5. Voted yes on 12 miles and no on six-plus-six. Therefore strong support for wide limits and strong opposition to narrow limits.
6. Voted no or abstained on both proposals therefore rejecting the framework of the debate.

(3) = territorial sea claim up to 1960

__ = fisheries claim over 12 miles

\- = landlocked state

Note: Categories 2 and 4 of the above are subject to alternate interpretations. For example, El Salvador's votes probably represented a position more in line with 5 than with 2. El Salvador's vote against a 12 mile limit reflected its own 200 mile preference and its abstention on six-plus-six probably reflected factors such as US pressure. Similarly, Japan's abstain vote on six-plus-six indicates its strong support for three miles, and so it should properly be placed in column 1.

and this created a polarization between the great majority of Western states on the one hand and an almost solid bloc of Arab and East European states on the other. The Latin American and Afro-Asian states were almost evenly divided between the two positions, again largely on the basis of their existing claims. This configuration represented a mix between the typical cold war alignments of the time, and the genuine differences of interest based on the issue at hand.

Although the six-plus-six and 12-mile positions did not, at first sight, appear to be far apart, they represented very different approaches to the law of the sea problem. The superficial similarity between the two was a measure of how influential the technological, functional, and political pressures in favor of wider coastal state jurisdiction had been. Most of the states supporting the six-plus-six proposal favored the traditional three-mile limit, and had fielded the six-plus-six proposal as a compromise between themselves and the large group of six- and 12-mile states. Their concern was that wide territorial sea limits would interfere with freedom of navigation, both naval and commercial, particularly by closing many important international straits. To avoid this they were prepared to concede functional jurisdiction to the coastal state, like that over the continental shelf, for fishing out to 12 miles, and also, through the idea of a contiguous zone, for customs, fiscal, immigration, and sanitary regulation.

The 12-mile states, by contrast, represented coastal rather than maritime interests. Territoriality and security were important motives behind their stand, and many of them, particularly the Latin Americans, had important interests in extensive claims, both territorial and functional, beyond 12 miles. In addition, the 12-mile claims of some of the Arab states must be viewed in terms of the

---

the 1958 conference, which called for a 12-mile maximum for the territorial sea along with a 12-mile fishing zone for those states claiming a territorial sea less than 12 miles. Since Guinea, Ethiopia, and Sudan did not vote on this, their position was taken from their sponsorship of a similar proposal (L.2 rev.1) in the committee at the 1960 conference. For six-plus-six, the data were derived from the United States-Canada proposal (L.11) in the plenary session of the 1960 conference. If no vote was available, as for Lebanon, that state's vote on the U.S. six-plus-six proposal in the plenary session of the 1958 conference was used. These two votes represent the best roll call data available on the two positions in that the majorities in favor on both were the largest mustered in support of the respective positions.

Arab-Israeli conflict where they served as a threat to Israel's access to the Indian Ocean. The fact that these wide versus narrow alignments coincided with cold war divisions meant that the deep divisions between the maritime and coastal state points of view were rendered even more complex, and that consequently the already limited possibilities for compromise at the two UN conferences were reduced to near insignificance. The existence of a superpower in the 12-mile camp undoubtedly made it much easier for the group of (mostly developing) countries that favored wider jurisdiction for the coastal state to resist the considerable pressure to accept the six-plus-six compromise.

The two conferences as a whole were characterized by alignments based on coastal interests—the desire to extend the limits and type of coastal state control over resources and activities off the coast—in conflict with maritime interests—the desire to preserve maximum freedom of the seas for all users. In some instances the two interests were in conflict within a single state (for example, in Britain), and on some issues compromises were found that enabled potential conflicts between the two types of interest to be bypassed (for example, on the continental shelf). As Table 2.3 shows, however, maritime interests were heavily identified with the Western states and their dependencies and allies, many of which were bound to the narrow limits system by both tradition and economic interest. Coastal state interests, by contrast, were heavily identified with the Soviet bloc and a miscellaneous group of developing countries, both of which had interests off their shores but neither of which were among the traditional fraternity of maritime states. The Soviet Union's commitment to a 12-mile territorial sea for fishing and security purposes was of long standing, and in this case it coincided conveniently with the Soviet desire, for cold war purposes, to build common ground with the developing countries. For these reasons the Soviet Union played an important role at the conferences in maintaining the opposition of the newly emergent coastal state interests to the long-established and well-entrenched maritime state interests. It must be noted, however, that by this time the developments that were to make the Soviet Union a major maritime state in the 1960s were already well under way, and that by 1960 her interest was confined to a 12-mile territorial sea and opposed to extensions beyond that limit.[13]

The two conferences were dominated by cold war divisions and maritime versus coastal state alignments; no other groups or alignments of importance occurred. Chile, Ecuador, and Peru worked closely together as a group, but were unable to attract much support from other Latin American states; they failed completely to make the Principles of Mexico a major issue at the conference. The

Peruvian attempt to introduce the principles received an unsympathetic response from most delegates, and the Peruvian delegation withdrew its proposal claiming that the conference had failed to study adequately the technical, biological, and economic aspects of the law of the sea.[14] Their efforts in this regard are interesting principally because of the moderate interpretations of the Declaration of Santiago forthcoming from the three states, and the almost exclusive interest in fisheries, as opposed to the seabed, that their statements reveal.[15] The unity of the three states in the face of the indifference or opposition of the rest of the conference was further emphasized by their joint statement on April 27, 1958 reaffirming that their legislation remained in "full force", and resolving to continue the struggle for coastal state rights.[16]

The Western European states consulted together at the conferences, as did the Soviet bloc, but with the exception of Chile, Ecuador, and Peru, and the possible exception of the Arab states, the developing countries had no forum in which to try to coordinate their policies. For the most part the primary ties of the developing countries were to the United States, the Soviet Union, and the major European powers. The 12 landlocked states were aware of their common interest, and held a conference in February 1958 at which they worked out a group position on their right of free access to the sea. This question was the only one on which they worked together, and they did not make any attempt to oppose the extension of coastal state jurisdiction on the continental shelf or in the water column despite the fact that such extensions automatically excluded them. On these questions their alignments tended to relate primarily to cold war politics.

The net result of the 1958 and 1960 Law of the Sea Conferences was the four Geneva Conventions, all of which resulted from the 1958 sessions. All four conventions passed by very large majorities, and taken together they represented a substantial codification of the existing international law of the sea. They did not, however, settle the problem of limits for coastal state jurisdiction over fisheries or in the territorial sea, and they left unresolved the definition of inner and outer boundaries of the continental shelf. Thus, while the Geneva Conventions clarified many of the less controversial aspects of the law of the sea, such as piracy and drawing of baselines, and facilitated the delimitation of marine boundaries between neighboring and opposite states, they did not close the book on law of the sea problems. The many volumes of legal analysis of the ambiguities in the Continental Shelf Convention alone illustrate this fact,[17] as does the disappointing record of ratifications to the conventions. In addition, the failure of the conference to pass a compulsory settlement of disputes provision indicated the extent of dissatisfaction with some of the compromises underlying the superficial consensus.

Each of the conventions required 22 ratifications before it entered into force, but despite the fact that 45 states signed the Continental Shelf Convention in 1958, it did not come into force until June 1964, six years after it was opened. The ratifications of the other conventions proceeded at much the same rate. That on fishing did not enter into force until 1966, and even by August 1972 only 50 states—fewer than two-thirds of those present at the conferences and a number representing only one-third of the number of extant states—had ratified the Continental Shelf Convention. No Arab states had ratified it, and only nine African states, six Asian states, and nine Latin American states, all from the Central America-Caribbean region, were among the 50. The instability inherent in this lack of enthusiasm for the compromise was pointedly underlined in January 1960 by the manned descent of the bathysphere Trieste 35,800 feet to the floor of the Challenger Deep—the deepest part of the seabed known to man. The conventions on the high seas and the continental shelf were not even in force before new technological developments began to undermine the assumptions on which they were based.

NOTES

1. The three-mile figure is based on S. Houston Lay, Robin Churchill, and Myron Nordquist, New Directions in the Law of the Sea (Dobbs Ferry: Oceana, 1973), p. 873. The other figures are the author's, calculated from a variety of sources.

2. M. W. Mouton, "Recent Developments in the Technology of Exploiting the Mineral Resources of the Continental Shelf," UN Doc. A/CONF.13/25 (1958). Other sources on this subject include Arvid Pardo, "Who Will Control the Seabed?", Foreign Affairs 47, no. 1 (October 1968), p. 125; Ernest Katin, op. cit., pp. 56-57; and H. R. Berryhill, Jr., The Worldwide Search for Petroleum Offshore—A Status Report for the Quarter Century 1947-72, U.S. Department of the Interior, Geological Survey Circular 694, 1974.

3. A/CONF.13/C.4/L.1, reprinted in A/CONF.13/42. For the sake of brevity, and since the discussion here falls within the context of only one committee at a time, all proposals will be indicated by "L" number only. All votes will be given in the order "support-oppose-abstain." The same procedures will be used for subsequent conferences described in later chapters.

4. The roll call for this vote is listed in A/CONF.13/42, p. 47.

5. The roll call for this vote is listed in A/CONF.13/SR.8, p. 7.

6. Marjorie M. Whiteman, "Conference on the Law of the Sea–Convention on the Continental Shelf," American Journal of International Law 52, no. 4 (Fall 1958), p. 637.

7. A. H. Dean, "The Geneva Conference on the Law of the Sea: What Was Accomplished," ibid., p. 619.

8. Whiteman, op. cit., p. 659. All three opposed the exploitation criterion, the inclusion of sedentary species in continental shelf resources, and the lack of a dispute settlement clause in the convention. France, Norway, and Sweden were among the abstainers.

9. Juraj Andrassy, International Law and the Resources of the Sea (New York: Columbia University Press, 1970), p. 132.

10. A/CONF.13/42, p. 34.

11. On this subject generally, see the study by the Secretary-General, UN Doc. A/AC.135/19 and Adds. 1 and 2 (1968), pp. 14-18.

12. The territorial sea data from the 1958 and 1960 conferences is found in UN Docs. A/CONF.13/C.1/L.1 to L.168 and SR.1 to SR.66, and A/CONF.19/8.

13. On this subject generally, see William E. Butler, The Soviet Union and the Law of the Sea (Baltimore: Johns Hopkins Press, 1971), pp. 42-45.

14. A/CONF.13/C.1/L.133 revs. 1 and 2, and C.1/SR.56, p. 6. See also Douglas M. Johnston and Edgar Gold, The Economic Zone in the Law of the Sea, Law of the Sea Institute Occasional Paper no. 17 (Kingston, University of Rhode Island Press, 1973), pp. 3-4.

15. Inter-American Juridical Committee, OEA/Ser.I/VI.2 CIJ-80 (Washington, D.C.: Pan American Union, 1966), p. 41.

16. A/CONF.13/L.50. An illuminating U.S. view of the motives of various states at the conference, including Chile, Ecuador, and Peru, can be found in Dean, op. cit.

17. See Shigeru Oda, "Proposals for Revising the Convention on the Continental Shelf," Columbia Journal of Transnational Law 7, no. 1 (1968); Myres McDougal and W. T. Burke, The Public Order of the Oceans (New Haven: Yale University Press, 1962); Barry B. L. Auguste, The Continental Shelf (Geneva: Librairie E. Droz, 1960); W. R. Neblett, "The 1958 Conference on the Law of the Sea: What Was Accomplished," in Lewis M. Alexander (ed.), The Law of the Sea: Offshore Boundaries and Zones (Columbus: Ohio State University Press, 1967); Edward D. Brown, The Legal Regime of Hydrospace (London: Stevens, 1971); and B. H. Oxman, "The Preparation of Article 1 of the Convention on the Continental Shelf," Journal of Maritime Law and Commerce 3 (1971-72).

CHAPTER

3

THE UNDERMINING OF THE GENEVA CONVENTIONS, 1961-67

Ocean politics in general, and seabed politics in particular, between 1961 and 1967 were characterized by four developments that tended to undermine the Geneva Conventions as a stable basis for the law of the sea. First, as noted in the previous chapter, was the fact that the conventions were slow to come into force, and that they did not attract the ratifications of a majority of states in the international system. Second was the influx of dozens of new states, mostly African, into the international system as a result of the last great round of decolonization. All of these new members were developing countries, and few of them felt much commitment to a system of international law that had been erected without their participation and reflected the dominance of former colonial maritime powers. Third was the growth of the Soviet Union to full maritime state status and the consequent shift of its interests to a position more in line with that of the traditional maritime powers. Fourth was the continued expansion in use of the oceans, combined with an intensification, especially in the United States, of research and development in marine science and technology. The period was also marked by a natural lull in international activity on the law of the sea following the concentrated effort of the 1958 and 1960 conferences. The initiative in ocean affairs was once again in the hands of individual states.

THE DEVELOPING COUNTRIES

The primary activity among the developing countries on the law of the sea during this period was the making of new claims. Many of these claims were made by newly independent countries, but a significant number were extensions of jurisdiction by older countries. No significant regional activity occurred among the developing countries during these years, but new organizations were in the making that were to be important in the later 1960s.

The momentum of decolonization resulted in a tendency for the developing countries to weaken their ties with the ex-colonial powers and strengthen their regional ties. It was in this period that the idea of the developing countries as a group with common interests became widespread, and found concrete expression in the Group of 77. The Group of 77 was created in the context of the first UN Conference on Trade and Development (UNCTAD) in 1964, and quickly became the instrument through which the developing countries tried to mobilize their most easily realized form of power: large numbers of votes at international gatherings. Voting power was, along with individual and regional claims, the biggest lever available to the developing countries against the wealth and technology of the developed states. The Group of 77 took upon itself the monumental task of coordinating the diverse interests of the developing countries in international forums. Although this development had little influence on law of the sea matters in the period under discussion, the habits and institutions of cooperation among the developing countries which were created at this time were to play a crucial role when the law of the sea was once again taken up at the international level.

In Latin America, the Inter-American Juridical Committee continued its efforts to find a common policy on the law of the sea for the continent. The Committee produced a report in July 1965 in which it recognized a three- to 12-mile territorial sea as acceptable under customary international law, and recommended a regional agreement based on the 12-mile maximum, but with the proviso that such an agreement would not affect states' rights to "adjacent zone" claims.[1] This work, however, had nothing like the impact of the IACJ's formulations in the early and mid-1950s, and does not appear to have had any tangible result.

Much more important in the Latin American context were the 200-mile territorial sea claims made by Argentina (1966), Ecuador (1966), and Panama (1967). These claims, all by states with existing linked continental shelf and fisheries claims beyond 12 miles, appeared to be leading a new Latin American assault on the dominance of narrow limits. The Continental Shelf Convention had severely weakened the strategy (using continental shelf claims as the basis for extended claims in the water column) that had been the basis of the Latin American coastal state offensive in the 1940s and 1950s. While the law of the sea conferences had closed this avenue, however, they had confirmed the inability of the international community to agree on a maximum breadth for the territorial sea, and had therefore left a loophole for unilateral action. The leading Latin American coastal state advocates were thus once again exploiting the weakest points of international law in order to advance their own interest in wider coastal state jurisdiction.

The Arab and Asian states contributed their share to the flow of new continental shelf and territorial sea claims in this period, but all of their claims were within the respective norms of 200 meters plus exploitation and 12 miles established by the UN conferences. The Persian Gulf states were occupied with a massive increase in offshore oil production (Saudi Arabia alone recorded a fourfold increase in annual production from 63.5 to 245.6 million barrels between 1960 and 1967), and it is noteworthy that none of the states bordering that very special body of water ratified the Continental Shelf Convention. There were no signs of regional activity on the law of the sea in either of these groups except for the slow continuation of bilateral delimitation agreements in the Persian Gulf.

For the African states, the key phenomena of this period were the huge influx of new states and the founding of a new regional organization, the Organization of African Unity (OAU), in 1963. Not counting the northern tier of Arab states, 30 new countries joined the African group between 1960 and 1969, changing it in less than a decade from the smallest to the largest regional group in international politics. Although the OAU did not interest itself in the law of the sea at this stage, its formation, like that of the Group of 77, was to become significant later in the decade.

The claims made by developing countries during this period completely changed the balance of alignment on the territorial sea from what it had been in 1958-60. Only four developing countries chose the six-plus-six format, while 19 claimed 12 miles, thereby changing the 12-mile position from that of a beleaguered minority to that of a group equal to, if not larger than, the narrow limit states. Perhaps most significant, though, was the 130-mile claim of Guinea (1964) and the 18-mile claim of Cameroon (1967). These two claims, though not on a par with the 200-mile territorial seas of the four Latin American countries, constituted an important breach in the 12-mile maximum limit. They made such claims an international rather than purely a Latin American phenomenon, and thereby added considerable strength and credibility to the Latin American initiative. Only eight developing countries made claims to the continental shelf in this period, and all but one of these, a 200-mile zone claim by Nicaragua, were within the terms of the Continental Shelf Convention.

THE DEVELOPED COUNTRIES

The primary relevant activity of the developed countries during this period was not claims, of which there were few, but increases in the use and capability of—and the attention paid to—marine technology. This was not a new development, but an acceleration or culmination of trends already visible in the 1950s.

During the 1960s, the Soviet Union firmly established itself as a major maritime power. Programs started in the 1950s resulted in a massive expansion of Soviet high seas fishing and oceanographic capability in the early 1960s. These were followed by equally impressive programs that came to fruition later in the decade to increase naval and commercial shipping capability. Soviet trawlers appeared off the east coast of North America in 1961 and off the west coast in 1966, and their catch quickly placed the Soviet Union in the top ranks of distant water fishing states along with Japan. The Soviets built large numbers of research vessels and created the educational infrastructure to produce the scientists and technicians necessary to support expanded programs. One author estimated that in the mid-1960s the scale of the Soviet Union's efforts in oceanographic research outpaced that of Britain, France, and Japan, and drew level with that of the United States.[2] The Soviet move to the oceans can be explained by, among other things, the country's need for fish as a source of protein to make up for deficiencies in meat production and the great increase in the military importance of the ocean that resulted from the development of submarine-launched ballistic missiles as secure second-strike strategic weapons.

These Soviet moves encouraged, and were encouraged by, developments in the United States, so that the two countries competed in an "ocean race" that was in many respects a smaller and less publicized parallel to the space race. There was no pressure in the United States to expand distant water fishing, and so American activity in the oceans was led by military and mining interests.

Attempts to get a national oceans program of some sort going in the United States dated back to 1959 when the National Academy of Sciences' Committee on Oceanography, responding to an initiative by the Office of Naval Research and Development, issued a report calling for a ten-year program to develop appropriate scientific knowledge, trained manpower, and technology. Eisenhower responded simply by setting up an Inter-Agency Committee on Oceanography, so it was not until the Kennedy administration that any effort was made to implement a long-range program. Kennedy submitted an ambitious program to Congress, but it fell to Johnson to actually get the project under way. He was greatly assisted in this task by the loss of the nuclear submarine *Thresher* in April 1963.

The Thresher sank in deep waters, and the ensuing search for the wreckage—a matter of great importance to the military in determining the cause of the loss—pointed out the severe limitations of existing deep water technology. Not only was the search process time-consuming and cumbersome, but it was also evident that the Navy possessed no submarine rescue capability beyond relatively shallow depths. Given the new strategic importance of submarines this was unacceptable, and funds quickly became available for deep water technology research and development. In 1964 projects for development of deep submergence systems and rescue vehicles were begun. The subsequent loss of nuclear bombs off the coast of Spain in 1966, and the sinking of another nuclear submarine, the Scorpion, in 1968 gave similar if less dramatic boosts to American deep water technology development.

The U.S. Navy was also involved in the Sealab project, which from 1964 on resulted in a series of experiments in sustained occupation of underwater dwellings. Two men spent 49 hours at 430 feet, and four men spent 11 days at 192 feet in the summer of 1964. The following summer similar experiments were performed in an improved sealab, and a man spent 30 days at 205 feet. These projects were significant not only because of military involvement and funding, but also because they brought the American aerospace industry into the field of marine technology. The large-scale involvement of these companies pushed the field well beyond the limits of the oceanographic institutes such as Scripps and Woods Hole that were its traditional base. Although a great deal of important technological development continued to come out of the oceanographic institutes, particularly in the field of small, deep-diving submarines, the intervention of government, military, and commercial interests motivated by a desire to fulfill particular goals, resulted in a rapid acceleration of marine technology development. Between 1956 and 1968, U.S. government spending on oceanographic research and development rose from 25 million to 448 million dollars, and over 60 research ships were built, converted, or authorized.[3]

The nation's long-term commitment to an expanded concern with the oceans was confirmed by the Marine Resources and Engineering Development Act and the National Sea Grant Colleges and Programs Act, both of 1966. These acts, which placed the government's commitment on a firm base, were an obvious response to the needs created by the various ocean projects and to the parallel actions of the Soviet Union. The acts declared a long-range program in marine science, and put emphasis on, among other things, accelerated development of the resources of the marine environment, encouragement of private enterprise in the field, preservation of the role of the United States as leader in marine science and resource development, development of all aspects of marine technology, and development of the necessary skilled manpower through the establishment of appropriate colleges and education programs.[4]

This whole process of creating a national commitment to an ocean program was greatly assisted by the continued rapid expansion of the offshore oil industry, and the increasing national reliance on its products, and by the interest that was springing up in the wake of new technology in the possibility of recovering copper, nickel, cobalt, and manganese, all of which the United States had to import in large quantities from nodules on the deep seabed.

Taken together, these factors created a massive momentum in the drive to increase American capabilities in marine science and technology. Much of this drive was oriented towards improved capabilities in deep water operations for both military and commercial purposes, and as such it opened up possibilities that had not been considered relevant in the deliberations of the 1958 and 1960 conferences. In particular, it opened up prospects for extensive military and hard mineral mining uses of the seabed that had scarcely been thought of before, and that were certainly outside the context of the oil-dominated perceptions of seabed utility that had dominated previous debates on the subject.

By the mid-1960s the conflict potential of these new developments was becoming clear. International law on the seabed beyond the continental shelf was ambiguous and therefore controversial, and the new technology appeared to be on the brink of nullifying the concept of "inability to occupy" that underlay the res communis interpretation of the high seas, including the seabed, beyond national jurisdiction. The possibility of occupation raised the question of claim to ownership, which in turn raised the prospect of conflict with other states. President Lyndon B. Johnson recognized this problem in a speech given in July 1966, after the passage of the Marine Resources and Engineering Development Act:

> Under no circumstances, we believe, must we ever allow the prospect of rich harvest and mineral wealth to create a new form of colonial competition among the maritime nations. We must be careful to avoid a race to grab and hold the lands under the high seas. We must ensure that the deep seas and the ocean bottoms are, and remain, the legacy of all human beings.[5]

This statement can be seen as the opening shot in the great debate in which the reopening of the seabed question was to lead to the reopening of the whole law of the sea. It is ironic that this occurred in the same year in which the last of the Geneva Conventions came into force.

During this period, the other developed countries also upgraded their interest in marine science and technology, but not on the same scale or with the same intensity as the United States and the Soviet Union. Most of them participated in the large-scale oceanographic research programs that arose out of the 1957-58 International Geophysical Year, but were less motivated by direct economic and military interests than the two superpowers.

Ten European countries plus Australia and New Zealand made continental shelf claims in line with the 1958 convention, and delimitation agreements between West Germany and the Netherlands (1964), Britain and Norway (1965), West Germany and Denmark (1965), Britain and the Netherlands (1965), Denmark and Norway (1965), Britain and Denmark (1966), and Denmark and the Netherlands (1966) resulted in the almost complete apportionment and allocation of the North Sea shelf among the bordering states. West Germany, Denmark, and the Netherlands could not agree on the overall delimitation of their shelf boundaries, however, and in 1967 they submitted their dispute to the International Court of Justice. At this time the North Sea was becoming increasingly important because of discoveries of oil and gas under the seabed. Discovery of gas in the northern part of the Netherlands in 1959 led to exploration of the adjacent seabed areas, which resulted in the first discovery of North Sea gas in 1965. Other gas and oil discoveries followed, creating the beginning of the ongoing North Sea boom, and by 1967 the first seabed gas was being piped into Britain.

Canada was one of the few countries to continue multilateral political activity on the law of the sea following the 1960 conference. After the narrow defeat of the six-plus-six compromise formula for the territorial sea at the 1960 conference, Canada led a campaign, along with Britain, Australia and others, to get agreement among all the states that supported the formula. By 1962 the idea had received support from 40 states, but many of them had made their adherence conditional on the participation of the United States. The Americans had reverted to a three-mile position after 1960, and when Canada presented them with the results of the campaign in 1962, they refused to participate. Their rejection torpedoed the whole effort, and Canada subsequently embarked on a sequence of unilateral claims to extend its maritime jurisdiction.[6]

DEVELOPMENTS IN MARINE TECHNOLOGY AND OCEAN ACTIVITY, 1961-67

Soviet and U.S. ocean activities were of course only a part, albeit a large part, of ocean activities worldwide. The increase in the scale of Soviet activity, and in the technological virtuosity of U.S. activity, can, however, be taken as symbolic of the global situation.

The massive expansion of fishing and shipping activities that went on all through the 1950s continued unabated in the 1960s. The shipping industry continued to enjoy an 8 percent annual increase in the tonnage of world sea trade, and began to shift to using much larger vessels for bulk commodities such as oil. The industry was one of the major beneficiaries of the increased trade in oil, but also found itself coming under increased pressure as a result of rising sensitivity to oil pollution. The wreck of the Torrey Canyon in the English Channel in March 1967 brought the whole supertanker pollution problem glaringly into the public spotlight.

Fishing continued to show a steep rise in the tonnage of the annual catch. The 1960 catch of 40 million metric tons was nearly double the 1950 total, and by 1967 had risen to 61.1 million metric tons, a rise of almost 300 percent from 1950. But the increase in scale brought with it an increasing variety and intensity of problems. Unrestrained competition for an essentially free resource resulted in the worldwide industry becoming highly overcapitalized, and consequently in excessive competition for, and overfishing of, the more desirable stocks. In response to this, by 1968 well over 50 states claimed fishing jurisdiction out to 12 miles, and at least another dozen states had claims between 12 and 200 miles.[7]

The use of the ocean as an ultimate dumping ground for sewage and industrial and nuclear wastes continued to expand rapidly through these years, and in this field also problems of saturation and hazardous use began to arise more frequently. The U.S. Atomic Energy Commission, among others, regularly dumped thousands of tons of nuclear wastes per year into the Atlantic Ocean from 1946 onward, but ceased to do so after 1968 because of rising concern over the possible consequences if such highly poisonous material did not stay on the deep seabed.

By this time the offshore oil industry was undergoing massive expansion. After 1960, the industry broke away from its reliance on U.S. and Persian Gulf fields and quickly became a worldwide phenomenon. Although expansion of offshore activity in the U.S. and Middle Eastern fields contined at an astounding rate—one author quoted U.S. government sources estimating a fourfold increase in production from U.S. offshore fields and a tenfold increase from Middle Eastern fields between 1960 and 1967[8]—the most important element of the industry's expansion for seabed politics was the spread of oil prospecting and production to the shelves of many coastal states. Between 1960 and 1968 offshore oil and gas production spread to over 20 countries, including Iran, Indonesia, Australia,

Mexico, Trinidad and Tobago, Nigeria, Brazil, Gabon, Britain, Peru, the United Arab emirates, and Malaysia, while offshore exploration spread to over 50 states.[9] This precipitate expansion raised hopes of offshore oil wealth in almost every coastal state, and enormously boosted the awareness of coastal state interests in new and old countries alike. Offshore oil was valuable, nonrenewable, dangerous (because of pollution hazards), and apparently widely distributed, and the prospect of it brought the continental shelf back into prominence in seabed politics.

The spread of the offshore oil industry was encouraged by many factors including high demand for its product, improved understanding of seabed geology resulting from extensive oceanographic work, and continued improvements in technology. Production had followed exploration down to 300 feet, while exploration had progressed to waters over 1,000 feet deep. The industry's boom is suggested by one author's estimate that investment was over 1 billion dollars per year, and growth over 18 percent per annum;[10] in 1967 alone, over 500 million dollars in bonuses were taken for leases on the shelf off Louisiana. By 1967 offshore wells were accounting for 12 percent of U.S. oil production and approximately 8 percent of world oil production. The high risks attending the boom were indicated by the doubling of insurance rates on big ocean drilling rigs in 1966 because of the high loss rate in storms.[11] By the mid-1960s offshore oil was well on the way to rivalling fishing as a interest of coastal states.

By contrast to the oil industry, the offshore hard minerals industry remained of interest only to a relatively small and very mixed group of states. Major operations continued to be confined to shallow waters, and occurred off only a dozen countries, including Australia (heavy mineral sands), Japan (iron sands), Iceland (lime shells), Britain (tin sands, sand, and gravel), the United States (heavy mineral sands, sulphur, lime shells, sand, and gravel), Indonesia (tin sands), Thailand (tin sands), South Africa (diamonds), and Argentina, Denmark, Sweden, and Yugoslavia (all sand and gravel only). Exploration was under way on the shelves of perhaps an additional dozen countries, mostly for phosphorite and heavy metal sands or muds. Although the industry's technology had improved somewhat, particularly with the development of better hydraulic and air lift dredges, its maximum capability was only around 300 feet, and the majority of its operations were in depths much shallower than that. In terms of value, the offshore hard minerals industry in the United States produced about 50 million dollars worth of minerals in 1966 compared with a worldwide production value for the industry of between 150 and 200 million dollars. These figures amounted to about one-twentieth of the comparative figures for the offshore oil industry, and even that proportion was shrinking rapidly with the much faster growth in offshore oil.[12] Bulk commodities like sand and gravel remained the backbone of the industry.

The one exception to the generally undramatic outlook of the offshore hard minerals industry was the small but growing excitement in the United States about the prospects for mining manganese nodules on the deep seabed. Manganese nodules had been known as a geological curiosity for nearly a century, but economic interest in them began to quicken in the early 1960s as a result of new work pioneered by U.S. oceanographic institutes. In 1957-58 an expedition from the Scripps Institute recovered a haul of nodules containing significant amounts of cobalt off the seabed near Tahiti, and succeeding oceanographic expeditions found nodules containing useful amounts of copper and nickel in other parts of the Pacific. The giant U.S. mining companies soon began to pick up an interest in the subject. International Nickel began to study the scientific work on nodules; Deepsea Ventures (Tenneco) in 1962 commenced a program of prospecting, technological research, and development as well as economic feasibility studies; and Kennecott Copper began studies in 1965 leading to an initial prospecting cruise in 1967.[13]

A landmark in this development was the 1965 publication of John Mero's book The Mineral Resources of the Sea. The book contained an enthusiastic appraisal of the oceans' mineral wealth in general, and laid particular stress on the potential of the manganese nodules. It brought the nodule question into the public eye, and can be seen as the major trigger of the great debate over the potential of nodule mining that was to rage in U.S. academic and commercial circles in the later 1960s. Mero himself backed up his beliefs by founding Ocean Resources Inc. in 1965. The holding of the first annual conference of the Law of the Sea Institute at the University of Rhode Island in June 1966 can also be taken as a landmark in the opening of this debate. The stakes were seen as being the more than 800 million dollars that the United States spent annually in importing manganese, copper, cobalt, and nickel.

The debate itself is more appropriately the subject of a subsequent chapter. It was only becoming significant toward the end of the period under discussion here, but even at this early stage it was to have a major influence on propelling seabed, and later ocean, politics back into the international arena. In the American context it should be noted that the debate coincided with, and was an increasingly important part of, the long-term national commitment to increased use of ocean resources.

By the end of this relatively short period, then, both the context and the issues of seabed politics were notably different from what they had been in 1958-60. The influx of new states and the creation of new groupings had substantially altered the international political context of ocean politics. Technology had begun to open up prospects on the deep seabed, and had encouraged coastal state interests by making much more widespread the hope for continental shelf oil. The continued rapid growth in all ocean uses had exacerbated the problems already visible in the 1950s, and had thereby both created more pressure on the regime embodied in the Geneva Conventions, and reinforced the awareness of coastal interests in the by then much more numerous group of coastal states. By 1967, the cumulative effect of legal, political, technological, and economic actions by states had reached a level sufficient to propel seabed politics back on to the agenda of the United Nations.

NOTES

1. Inter-American Juridical Committee, OEA/Ser.I/VI.2 CIJ-80 (Washington, D.C.: Pan American Union, 1966).

2. M. D. Shulman, "The Soviet Turn to the Sea," in Edward A. Gullion (ed.), Uses of the Seas (Englewood Cliffs, N.J.: Prentice-Hall, 1968), pp. 145-46. See also W. E. Butler, "Some Recent Developments in Soviet Maritime Law," in Lewis M. Alexander (ed.), The Law of the Sea: The United Nations and Ocean Management (Kingston: University of Rhode Island Press, 1971), pp. 375-86; and William Butler, The Soviet Union and the Law of the Sea (Baltimore: Johns Hopkins Press, 1971).

3. Arvid Pardo, "Who Will Control the Seabed?", Foreign Affairs 47, no. 1 (October 1968), p. 129; and P. M. Fye, A. E. Maxwell, K. O. Emery, and B. H. Ketchum, "Ocean Science and Marine Resources," in Gullion, op. cit., p. 24. On this subject generally, see also Norman J. Padelford, Public Policy for the Seas (Cambridge: MIT Press, 1970), p. 8; Brenda Horsfield and Peter Bennet Stone, The Great Ocean Business (London: Hodder and Stoughton, 1972), pp. 174-75, 185, 234; and Jackson M. Tomsky, "Sealab Man-In-The-Sea: Future Phases from 1969," in Progress Into the Sea (Washington: Marine Technology Society, 1969), pp. 267-69. The U.S. Navy's involvement in Sealab conjured up fears in some quarters of permanent military bases on the seabed.

4. Public Laws 89-454 and 89-688, 89th Congress, respectively June 17 and October 15, 1966. Reprinted in Padelford, op. cit.

5. Quoted in Louis Henkin, "Changing Law for the Changing Seas," in Gullion, op. cit., p. 82. See also the New York Times, July 14, 1966, p. 10.

6. L. H. J. Legault, "Maritime Claims," in R. St. J. Macdonald, G. L. Morris, and D. M. Johnston (eds.), *Canadian Perspectives on International Law and Organization* (Toronto: University of Toronto Press, 1974), pp. 382-83; and Ann L. Hollick, "Canadian-American Relations: Law of the Sea," *International Organization* 28, no. 4 (Autumn 1974), pp. 755-80.

7. Fye et al., op. cit., p. 65, list 63 states claiming fisheries jurisdiction to 12 miles and 13 claiming beyond 12 miles. Only 13 states are listed as maintaining three miles and ten as maintaining six miles. For fishing statistics see *FAO Yearbooks of Fishery Statistics*. On shipping see Robert L. Friedheim, "International Organizations and the Uses of the Ocean," in Robert S. Jordan (ed.), *Multinational Cooperation* (New York: Oxford University Press, 1972), pp. 224-26.

8. Gerard J. Mangone, *The United Nations, International Law, and the Bed of the Seas* (Newark: University of Delaware Press, 1972), p. 19, quoting from the U.S. Commission on Marine Science and Technology, *Panel Reports* 2, USGPO, 1969.

9. U.S. Department of the Interior, *Summary Petroleum and Selected Mineral Statistics for 120 Countries, Including Offshore Areas*, Geological Survey Professional Paper 817, Washington: USGPO, 1973; and Louis Henkin, *Law for the Sea's Mineral Resources*, ISHA Monograph no. 1, New York, 1968, p. 3.

10. James A. Crutchfield, "Resources from the Sea," in English T. Saunders (ed.), *Ocean Resources and Public Policy* (Seattle: University of Washington Press, 1973), p. 119.

11. G. J. F. MacDonald, "An American Strategy for the Oceans," in Gullion, op. cit., p. 167.

12. On production see Fye, et al., op. cit., p. 33; Padelford, op. cit., pp. 191-92; Crutchfield, op. cit., p. 130; Willard Bascom, "Mining in the Sea," in Lewis M. Alexander (ed.), *The Law of the Sea: Offshore Boundaries and Zones* (Columbus: Ohio State University Press, 1967), pp. 160-65; and U.S. Senate, *Mineral Resources of the Deep Seabed*, Hearings before the Subcommittee on Minerals, Materials and Fuels of the Committee on Interior and Insular Affairs, 93rd Congress, first session on S.1134, May 17, June 14, 15, 18, and 19, 1973, Washington: USGPO, 1973, p. 161. On technology see Feng-hui Wang, *Mineral Resources of the Sea*, UN Doc. ST/ECA/125, 1970, pp. 37-43.

13. *Mineral Resources of the Deep Seabed*, ibid., pp. 202-05.

CHAPTER

4

THE REASSERTION OF INTERNATIONAL INTEREST IN THE SEABED, 1961-68

THE UN, THE PARDO PROPOSAL, AND THE AD HOC SEABED COMMITTEE

From 1966 onward, the General Assembly of the United Nations developed an increasing interest in the oceans in general and the seabed in particular. This involvement related directly to both the expanding U.S. and Soviet ocean programs, and to the heightened awareness of coastal state interests, especially among the developing countries. It was different in nature from the UN's first involvement with the law of the sea in the 1950s in that it took place in the more open and overtly political forum of the General Assembly rather than in the International Law Commission, and also because it was focused on an issue that was largely outside the bounds of both international law and state practice: the seabed beyond the continental shelf.

The United Nation's involvement with the oceans through its specialized agencies was of very long standing. The Food and Agriculture Organization (FAO), the World Meteorological Organization (WMO), the Intergovernmental Maritime Consultative Organization (IMCO), the International Atomic Energy Agency (IAEA), the Intergovernmental Oceanographic Commission (IOC), and many others were all heavily committed to the practical side of ocean activity, and their work load reflected the increasing level and variety of ocean use. The existence of these organizations ensured that the UN as a whole was aware of the rising tide of ocean problems, and also of the new drives in technology and economic interest emanating from the United States.

The first important sign of political interest in the subject, however, came from the Economic and Social Council (ECOSOC) in its Resolution 1112 in March 1966. The council observed that "the mineral and food resources of the sea beyond the continental shelf, excluding fish, constitute reserves of raw materials which are as yet not fully being utilized, and that the rational use of these resources to ensure optimum yield and minimum waste is of vital importance to all countries." It went on to request the Secretary-General "to make a survey of the present state of knowledge of these resources . . . and of the techniques for exploiting [them]."

Other than being the first sign of political interest in the subject, this resolution was noteworthy for three reasons. First, it indicated the remarkable speed with which awareness of developments in the United States filtered through to the political and economic side of the United Nations. The ECOSOC made this request several months before President Johnson publicly identified the seabed as a possible area of economic interest and international difficulties. Second, the resolution showed clearly that the developing countries as a group were aware of their common interests in the subject. It contained several references to the economic interest of the developing countries in these resources that, taken together, amounted to an implicit declaration that the developing countries were decidedly unhappy with the prospect of a few developed states using their technological advantage to monopolize a new and important body of resources in an area not clearly owned or controlled by anyone. Third, the resolution was the first example of developing countries as an organized group using the UN as a source of information that they could not easily obtain for themselves. In this way the facilities of the UN acted as a counterweight for the developing countries to the technological, economic, and experiential knowledge of the developed states.

The ECOSOC resolution was followed in December 1966 by Resolution 2172 in the General Assembly. This passed by 87 to 0 to 12, and as well as endorsing the ECOSOC resolution it requested an additional report from the Secretary-General on a comprehensive survey of marine science and technology activities worldwide, and on proposals for increasing international cooperation and education in the marine sciences. The concerns of the developing countries for their economic interests, and for the acquisition of information, were both evident again in this resolution. In addition, the General Assembly's resolution had a more clearly political intent than that of ECOSOC, in that the report it requested would necessarily reveal which countries had big vested interests in marine science and technology. Any such survey of activities would demonstrate the enormous concentration of marine science and technology activity in a handful of highly developed countries, and the almost total absence of such activity among the developing countries. It would provide the developing countries with insight into the national ocean programs of the United States and the Soviet Union, and would help to clarify the scope and direction of economic interest in the oceans.[1] These two resolutions established the needs and interests of the developing countries as an important factor in the new economic interest in the oceans, and also set the stage for a major United Nations role in the emerging international controversy on the law of the sea.

At this point a unique event occurred without which the UN's involvement in the ocean debate would probably have unfolded more slowly, and certainly unfolded in a markedly different way. On August 17, 1967 Ambassador Arvid Pardo of Malta undertook an independent initiative on behalf of his country by requesting the inclusion of this supplementary item on the agenda of the General Assembly:

> Declaration and treaty concerning the reservation for peaceful purposes of the seabed and of the ocean floor, underlying the seas beyond the limits of present national jurisdiction, and the use of their resources in the interests of mankind.[2]

Pardo's initiative was a bold addition to trends already under way. It resulted substantially from his own interest in the subject, and was made on behalf of Malta alone, not as a product of planning by any group of states.[3] It was designed to stimulate action on internationalization of the deep seabed before advancing technology made exploitation possible and resulted in a proliferation of ever larger national claims.[4] Although it failed to achieve this objective, it was nonetheless an extraordinary attempt to force history to change direction. It precipitated a flurry of activity at the United Nations, and set the framework for an important part of the discussion in the succeeding years.

Pardo began by describing the existing situation in terms of a profound clash between the interests of developed and developing countries:

> In view of rapid progress in the development of new techniques by technologically advanced countries, it is feared . . . that the sea-bed and the ocean floor, underlying the seas beyond present national jurisdiction, will become progressively and competitively subject to national appropriation and use. This is likely to result in the militari-

> zation of the accessible ocean floor through the establishment of fixed military installations and in the exploitation and depletion of resources of immense potential benefit to the world, for the national advantage of technologically developed countries.[5]

He proposed to resolve the dilemma by a declaration that the seabed and ocean floor were "a common heritage of mankind" and by a treaty containing principles to make this common heritage operational. These principles included nonappropriation by states, use of seabed revenues primarily to promote the development of poor countries, reservation of the area for peaceful purposes only, and creation of an international agency to assume jurisdiction over the seabed beyond national jurisdiction and to "regulate, supervise and control all activities thereon."[6]

Pardo's proposal was a key event in the history of seabed politics for a number of reasons. First, it separated the seabed from the general matrix of ocean issues and successfully established it as an important focus of concern. Second, it gave an immense momentum to the idea of the common heritage of mankind by linking it to the needs and interests of the developing countries. Third, and related, it set the issue squarely in the context of developed versus developing country interests, and thereby established, or greatly helped to establish, alignments on the issue that might otherwise have evolved in different, or at least in less stark, forms. Fourth, it raised the vision of a new type of international organization that would give economic benefits and political power to the developing countries as a means of restoring the balance between them and developed countries. Fifth, it raised a banner against technological imperialism, of both the economic and the military variety, around which the developing countries could rally with relative ease. And sixth, it placed the policy initiative regarding a regime for the deep seabed firmly within the framework of international organization before any states had made claims to that area.

The proposal was an extraordinary amalgam of shrewd analysis and political insight on the one hand, and crippling ambiguity, contradiction, and excessive optimism on the other. Its tone and content virtually assured it a place of pride among the objectives of the Group of 77, and therefore a powerful base in international politics. At the same time, however, its utterly uncompromising denunciation of developed country interests, particularly military interests, condemned it to become a political issue of the most bitterly divided kind, and therefore effectively ruined any chance it might have had for rapid adoption—which was crucial if it was to avoid being overtaken by the very technological and economic developments of which it warned. The whole idea of "peaceful purposes", for example, was hopelessly vague, and justified only by reference to unlikely military scenarios. It related to problems of definition that disarmament bodies had been unsuccessful in resolving despite years of effort, and it had the effect of alienating the developed countries, muddying the central economic issues of the question, and introducing essentially unresolvable disputes into the debate. Seen in this light, Pardo's proposal looks much more like a spirited, ideological proclamation on behalf of internationalism than an innovative compromise designed to attract the wide support necessary for quick implementation. It hastened the onset of both the conflict between the developed and developing countries over the deep seabed, and the dialogue aimed at creating a new international regime and machinery for the seabed beyond the limits of national jurisdiction. By stimulating these two conflicting trends in combination, Pardo's proposal did much to fashion the flawed mold in which future discussions of the subject were cast.

The immediate result of Pardo's proposal was the passage of Resolution 2340 on December 18, 1967 by 99 to 0 to 0. This resolution established a 35-state-ad hoc committee to study the peaceful uses of the seabed and the ocean floor beyond the limits of national jurisdiction (referred to hereafter as the Ad Hoc Seabed Committee or AHSBC). The wide support it attracted resulted from compromises worked out in the first committee of the General Assembly between the United States and others who favored establishing a permanent committee on the oceans, and the Soviet Union, which opposed founding a committee; and from a watering-down of Pardo's original wording so that "Declaration and treaty concerning" became "Examination of the question of" and most of Pardo's concrete proposals were not mentioned. These dilutions were made necessary by the fears of the Latin American states that the item would raise questions about the limits of national jurisdiction and by the concern of the maritime powers, especially the United States, that it would disrupt development of seabed mining.[7]

The resolution, however, was far from an empty gesture. It carried a strong implication that there was an area beyond the limits of national jurisdiction, and thereby helped refute the contention that the Continental Shelf Convention allowed states to claim out to the median line between themselves and their nearest transoceanic neighbors. Even more important was the mandate given to the AHSBC. Following tradition, the General Assembly instructed the AHSBC to prepare studies (to be submitted to the 1968 General Assembly) on various aspects of the question including scientific, technical, economic, and legal matters. In addition, it gave the committee a wide-ranging prescriptive function:

> [to indicate] practical means to promote international cooperation in the exploration, conservation and use of the sea-bed and the ocean floor, and the subsoil thereof, . . . and of their resources, having regard to the views expressed and the suggestions put forward by Member States during the consideration of this item at the twenty-second session of the General Assembly.

This opened the way for discussion of Pardo's proposals, and given the unanimous support of the mandate, amounted to a fairly potent achievement. The committee itself comprised a mixed group with an emphasis on the most generally influential states of each region. It contained 16 developed and 19 developing countries, and was notable for its proportional under-representation of landlocked states (two), shelflocked states (two) and other small-margin states (two). The proportional over-representation of developed states, both Soviet group and Western, made the AHSBC a much safer place than the General Assembly for them to discuss Pardo's proposals.

The AHSBC held six weeks of meetings during 1968, and the character of its work was very much determined by the nature of its mandate. In order to prepare factual studies the committee had to acquire more information, because to begin with it had only the two Secretary-General's reports resulting from the 1966 ECOSOC and General Assembly resolutions. These reports came out respectively in February and April 1968, and although they were thorough on questions of marine science, technology, and resources, they were not intended to cover economic, legal, and military aspects of the question. The situation was similar with regard to the prescriptive side of the committee's mandate. The guidance given by the General Assembly was quite open-ended, and the committee had no precedents or proposals other than Pardo's around which it could organize its work. There was, by definition, no state practice in the area, and neither the Continental Shelf Convention nor the High Seas Convention contained much that was applicable to the problem. The AHSBC thus began its work with neither sufficient information at its disposal nor a precisely defined task to fulfill.

To facilitate its work the committee created two working groups, one to deal with economic and technical aspects, and one to deal with legal matters. It retained for itself discussion of scientific aspects and of recommendations for promoting international cooperation. It also decided to conduct its work by consensus (where an item is not adopted until nobody opposes it) rather than by voting. This decision, taken over the strenuous objections of Tanzania, reflected the fear of the developed country delegates that in a voting situation they would become a permanent minority. Although the decision was made with relatively little fuss, it was to have a profound impact on the years of discussion that followed. It meant that the committee's work would proceed very slowly while each difficulty was resolved to the satisfaction of all members, but that the final result would represent a "true" agreement and not the dominance of one set of interests over another. It also raised the possibility that no agreement would ever be reached, and made it likely that any agreements that did emerge would be vaguely worded and general in nature. While the consensus rule by no means made the AHSBC apolitical, it greatly muted the political process available to the delegates and ensured the committee a future in which all options would be developed in exhaustive detail, and in which the movement toward compromise would be arduous and prolonged. The alternatives to consensus seemed to involve the likelihood of a quick conflict between the developed and developing countries, a paper victory for the latter, and a walkout by the former.

The AHSBC dissipated a good deal of its time in unfocused and frequently ill-informed debate, but nonetheless three themes emerged clearly in its work. First was the acquisition and absorption of large amounts of background information on the subject. In addition to the earlier reports on marine science, technology, and resources, the Secretariat produced a major report on the legal aspects of the question (A/AC.135/19 and Adds. 1 and 2) and several shorter reports on such aspects as international agreements concerning the area (135/10), relevant national legislation (135/11), economic implications of seabed mineral exploitation (135/14), effects of seabed mining on other uses of the ocean (135/15), and military uses of the ocean floor (135/28). IMCO and the IOC also submitted reports in response to requests by the AHSBC (135/4, 17 and 23). These reports enabled the AHSBC to fulfill the informational requirements of its mandate in its report to the General Assembly, even though many aspects of the item, particularly its economic implications, remained unclear. But more important, the reports enabled the committee to establish a foundation on which to base the prescriptive part of its work.

The second theme of the committee's work was the idea that a declaration of principles should be its goal in terms of fulfilling the prescriptive part of its mandate. The idea of a declaration of principles had been used by previous UN committees, and had the merit of providing an easily accepted and familiar focus for the AHSBC's deliberations. It had the additional advantage of requiring only a general level of agreement; this was most important given the temporal and procedural constraints on the committee's work and the controversial nature of its subject. India first broached the idea early in the committee's second session. Tanzania and

Libya supported it, and it eventually became accepted as the most promising way of giving form and direction to an otherwise wandering and diffuse debate. India, the United States, and a group (totalling 15 states) of all the developing countries on the committee except Senegal, Somalia, and Yugoslavia, all tabled draft declarations of principles. By the end of the AHSBC's sessions, debate and compromise had whittled the field down to two competing drafts, one virtually identical to the proposal of the 15 developing countries, and the other, proposed by Britain, supported by all the Western states.[8] The Soviet Union and most of its allies remained uncommitted, and the Committee as a whole was unable to reconcile the developed/developing split sufficiently to recommend a draft declaration of principles to the General Assembly.

The third theme in the AHSBC's work was the emergence of increasingly well-defined areas of disagreement. This theme was largely, but not wholly, related to the debate on the declaration of principles, and was accompanied by the identification of groups of states prepared to act together in support of common interests.

Perhaps the most important area of disagreement was Pardo's idea that the seabed beyond national jurisdiction should be declared a common heritage of mankind. Neither the U.S. nor the British drafts even mentioned the term "common heritage," while the draft of the developing countries gave it prominence as the first principle. At this stage, no systematic attempt had yet been made to define what the common heritage principle meant in practice, and thus most objections to it were on the grounds that it was legally meaningless. Alignments on this dispute were very definitely along developing versus developed lines.

Almost equally important, and much more visible in the committee's records, was the dispute over defining, and thereby delimiting, the area. Although no state questioned the assumption that there was an area beyond the limits of national jurisdiction, there was much dispute over how, and even whether, the area should be defined. The developing countries ignored this problem in their draft declaration of principles, while the developed states enshrined the need for a precisely defined boundary as their second principle. Despite this, it would be a mistake to assume that this dispute also divided states along purely developed/developing lines. For one thing, the question had several aspects, each of which encouraged different responses. Iceland, for example, reiterated its traditional desire to link the continental shelf to the superjacent waters, and this roused the United States, the Soviet Union, and Japan to their traditional opposition. More important was a trend among some countries with large margins, notably Argentina, Australia, Canada, the United States, and Italy, to assume that the area was the deep seabed beyond the limits of the geographical continental slope or margin. This trend was opposed vigorously by Sri Lanka, Tanzania, and Liberia, and the dispute foreshadowed a much more widespread confrontation on this point in later years.

Perhaps the most significant alignment to emerge on this issue, however, was the group of five Latin American states with 200-mile claims–Argentina, Chile, Ecuador, El Salvador, and Peru. This group signalled the existence of a substantial rift in the ranks of the developing countries between the Latin American states, many of which were concerned over the boundary implications of Pardo's proposal, and the African and Asian states, most of which embraced Pardo's proposal with unreserved enthusiasm. Its members strongly opposed both the idea of a moratorium on national claims suggested separately by Tanzania and Japan, and the proposal to establish a boundary for the area supported by most of the Western countries, Tanzania, and Yugoslavia. They feared not only that their claims would be compromised by any boundary agreed upon by a group of states with smaller claims, but also that the establishment of a boundary at such an early stage would stifle the chances of other states joining them by making large unilateral claims. Because of the AHSBC's decision to work by consensus, this group was able to block successfully any further progress on the question of defining a boundary for the area.

There was a predictable and fruitless dispute over the question of peaceful purposes in which the developed military powers defended their right to use submarines and detection devices against the demands of many developing countries for complete demilitarization of the area. With little possibility of meaningful compromise on this issue, this legacy of Pardo's proposal succeeded only in wasting time and embittering the atmosphere in the committee.

One other area of disagreement that also foreshadowed a much wider debate was the question of international machinery to put into effect the common heritage principle. The suggestion for such machinery came from Pardo's proposal, and resulted in a heated clash between Tanzania on the one hand and the Soviet Union and Bulgaria on the other. The Soviet Union wanted to exclude the topic from the AHSBC's agenda, and displayed its usual lack of enthusiasm for new international organizations. Tanzania took the question of international machinery as vital to fulfillment of the common heritage principle, and therefore an essential item of discussion. Like the other disputes, this one was not solved by the AHSBC before it concluded its meetings.[9]

The results of the AHSBC's year of work can only be described as inconclusive. Its discussions had been wide-ranging and rather general in nature, and had not reached agreement on any items of substance. It did recognize that there was a need to define a new regime for the area, and that the needs of the developing countries should be given special consideration in relation to exploitation of the area. There was general agreement that no state could claim or exercise sovereignty in the area, but this was effectively meaningless until some way was found around the impasse on defining boundaries. Lack of time, information, and a precise goal meant that very little detailed discussion took place on any subject.

The committee's sessions were, nonetheless, useful. The discussion not only served as an educational device for the less well-informed delegations, but also helped to define several key issues and to give purpose and direction to the whole seabed debate. Matters such as international cooperation in marine scientific research were discussed with surprisingly little disagreement compared with later years, and the committee accepted a Belgian proposal that a permanent seabed committee should be established to continue the work it had begun. Requests were made for more studies on areas where the debate had revealed deficiencies in existing sources of information, and although the committee's progress had been slower than many had hoped, there were grounds for optimism in that its work seemed to be moving in a constructive direction.

In the long term, however, the most important aspect of the AHSBC's work was the momentum it created toward reopening other law of the sea questions supposedly closed by the Geneva Conventions. The discussions on boundaries for the area reopened the question of continental shelf limits, and in some cases threatened to clash embarrassingly with territorial sea limits. Similarly, discussion of activities relating to the area raised questions about the suitability of existing high seas doctrines. The obvious interconnectedness of the seabed area with other aspects of the law of the sea led Libya and Iceland to call for a new conference on the subject. This proposal proved to be premature, but prophetic.

After the last meeting of the AHSBC on August 30, 1968, the seabed debate moved to the General Assembly. Since the Assembly worked by majority vote rather than by consensus, its records reveal more clearly the extent to which the developing countries were trying to use the UN as a means of counterbalancing the natural advantages of the developed states with respect to ocean issues. For example, ECOSOC in its resolutions 1380, 1381, and 1382 (August 2, 1968) asked the Secretariat to produce regular updates of its report on Resources in the Sea (E/4449), recommended a long-range ocean exploration program to the General Assembly, and asked UNESCO to study means of improving education in marine affairs. The Assembly adopted two resolutions on December 17. Resolution 2413, which passed 99 to 0 to 8, requested increased international cooperation in the rational exploitation of high seas fisheries with special consideration for the needs and interests of the developing countries. Resolution 2414, passed unanimously, reiterated the ECOSOC requests and in addition urged consideration of aid to states wanting to develop their continental shelf resources.

The debate on a declaration of principles moved to the General Assembly's First (Political) Committee, which had no more success than the AHSBC in bringing it to a conclusion.[10] A group of over 30 states, mostly Western, African, and Asian, favored adopting a set of principles during the 1968 session. These were opposed by a smaller group, including the Soviet Union and most of its allies, that felt the question needed more study, especially since there did not yet exist sufficient agreement to ensure wide support for any such declaration.

Since no agreement could be reached on a declaration of principles, the Assembly turned its attention to reconstituting the AHSBC so that it could continue to work on the problem. This was done in Resolution 2467A of December 21, 1968, which passed by a vote of 112 to 0 to 7 (the Soviet Union, the Ukrainian republic, the Byelorussian republic, Hungary, Cuba, Cambodia, and Equatorial Guinea abstaining), and which established a permanent Seabed Committee (hereafter SBC) of 42 members. The new SBC had much the same membership as its predecessor except that three developing countries had been dropped (Ecuador, Senegal, Somalia) and ten more developing countries added. Thus while the total of developed countries remained at 16, developing countries, comprised of eight African, five Asian, four Arab, and seven Latin American states plus Malta and Yugoslavia, totalled 26. Landlocked, shelflocked, and small-margin countries continued to be under-represented by only eight members. By contrast, two-thirds of the states with major interests in importing the four metals found in manganese nodules (nickel, copper, cobalt, manganese) were on the committee, as were all the states with commitments to creating a deep sea mining capability (with the exception of West Germany, which was not then a member of the UN). Fewer than one-third of the states with major interests as exporters of the four metals were members.

On the basis of the progress made by the AHSBC the General Assembly was able to give the SBC a more specific mandate. After making clear that any exploitation of the area should take into account the special interests and needs of the developing countries, the Assembly instructed the SBC to

> study the elaboration of the legal principles and norms which would promote international co-operation in the exploration and use of the sea-bed and the ocean floor,

> and the subsoil thereof, beyond the limits of national jurisdiction and to ensure the exploitation of their resources for the benefit of mankind, and the economic and other requirements which such a regime should satisfy in order to meet the interests of humanity as a whole.

This mandate was much more clearly oriented toward the elaboration of a regime for the area than was that given to the AHSBC. Like its predecessor, the SBC was also instructed to study the economic and technical aspects of the item, and to discuss the reservation of the area for peaceful purposes.

In addition to creating the SBC, the General Assembly included three other resolutions in the package of 2467. Two of these, 2467B and 2467D, were uncontroversial, involving respectively a request for a study on pollution arising from exploitation of the area, and a statement of support for the concept of an International Decade of Ocean Exploration (IDOE). Resolution 2467C, however, was very controversial, and its adoption by a vote of 85 to 9 to 25 was the first important public declaration of the growing dispute on the seabed issue. The resolution requested the Secretary-General to prepare a study on establishing international machinery for the promotion of the exploration and exploitation of the resources of the area, taking into special consideration the interests and needs of the developing countries. It further instructed the SBC to add this subject to its mandate.

The question of international authority had been raised in the AHSBC, and it will be recalled that the Soviet Union had opposed its inclusion on the committee's agenda. Outside the restraints of the SBC, however, the developing countries were free to wield their voting power. Since most of them agreed with Tanzania that international machinery was essential to the realization of the common heritage principle, it was not difficult for them to override the Soviet protest. Resolution 2467C was sponsored by 39 developing countries. All of the nine states opposing it were members of the Soviet group, and of the 25 states abstaining, 13 were Western, including all the major powers except Japan, and only 12 were developing countries, most of which had strong ties to either the Soviet Union or the Western powers. Of the 85 states supporting the resolution, only ten were Western: nine small European states and Japan. The voting power of the developing countries thus carried the day, and the vital subject of international machinery was injected firmly into the deliberations of the SBC.

The passage of 2467C marked substantial progress by the developing countries toward one of the goals originally laid down by Pardo. Although none of the resolutions in 2467 mentioned the common heritage principle, all of them were much more assertive about the "needs and interests" of the developing countries than resolution 2340 had been, and the inclusion of international machinery in the debate was a long step toward the common heritage view whether the principle was mentioned or not. Thus by judicious use of their voting power the developing countries were able to manipulate the mandate of the SBC in a direction favorable to their own objectives. They achieved this only at the cost of offending the Soviet Union, but at this stage the results of such an exercise in vote power were effective. The study requested was done, and the mandate was changed without any state withdrawing from the SBC. The voting alliance of the developing countries therefore had a substantial impact so long as the objectives of the vote were confined within the purview of the General Assembly.

Paradoxically, the interest of the developing countries in the deep seabed was heavily stimulated at this time by the debate on the subject occurring within the United States (discussed below). This debate was closely related to the internal politics of U.S. ocean interests, and consequently generated extravagant estimates of the potential wealth of the ocean floor. The idea of a share in the wealth was, at this stage, the principle component of the whole common heritage movement.

It was clear by this time that the Soviet Union was the country most opposed to the whole chain of events springing from Pardo's initiative. She had been conspicuously absent from the otherwise balanced list of 44 states that had sponsored resolution 2340 in 1967, and the Soviets had made it plain that they were unenthusiastic about the establishment of the AHSBC, even though they did not oppose the resolution in the vote. They preferred national research coordinated by existing UN maritime agencies to the prospect of a seabed authority of the type suggested by Pardo, and had made efforts in 1967 to move the IOC in this direction.[11] In February 1968, just before the first meeting of the AHSBC, the Soviet Union passed legislation to implement its 1960 ratification of the Continental Shelf Convention. This legislation amounted to a pointed defence of her substantial continental shelf interest against any encroachments by the AHSBC. It accepted the 200-meter plus exploitation definition of the convention, but in addition included troughs as part of the shelf, and specified that the convention applied only to the continental margin and not to the deep seabed beyond. It also interpreted the exploitation provision as meaning that the most advanced state set the limits for all others.[12]

The Soviets did not warm to the AHSBC all through 1968. Although they did not go so far as to align with the Western states, they opposed the common heritage principle and the idea of international machinery. Instead they stressed the need for more research in the oceans before any decisions could be made sensibly, and spent much time in the debate on peaceful purposes, where they had some hope of support from the developing countries. They did not support either draft of principles, and in the First Committee debate on the subject they emphasized the need for unanimous support of any principles that were adopted, and opposed the adoption of principles during the 1968 session. Their abstention on the vote to create a permanent SBC indicated continued dissatisfaction with the trend of events, and their vote against the study on international machinery indicated their considerable disaffection from the course favored by the developing countries.

The transition of the Soviet Union from staunch ally of the developing countries in 1958-60 to their most intransigent opponent in 1968 indicated not only the degree to which a burgeoning maritime capability had altered Soviet perceptions of the law of the sea, but also the degree to which the developing countries had shifted the primary UN alignments from an East/West to a developed/developing perspective. The shift still left the Soviet Union on the minority side, and the Soviet delegates simply continued their long tradition of staunch opposition to the will of an unsympathetic majority.

The position of the United States was somewhat different. Its status as a maritime power was of long standing, it did not share the Soviets' distrust of international organizations, and it was not used to finding itself in a permanent minority.

The United States supported the idea of a SBC, and participated actively in its work. Like the Soviets, the Americans were anxious to preserve the traditional freedoms of the high seas, and made this abundantly clear in their draft principles (A/AC.135/25). The original U.S. proposals gave short shrift to the common heritage idea—". . . dedication as feasible and practicable a portion of the value of the resources recovered from the deep ocean floor to international community purposes"—and concentrated instead on preserving a legal environment that would encourage private investment in seabed mining, and on defining a boundary for the area. This draft provided a detailed insight into U.S. thinking on the seabed issue at a very early stage in the negotiations. The U.S. delegation, however, quickly subordinated its own draft to the much shorter and more general draft offered as a compromise by the British, and the U.S. view did not return to prominence until 1970.

In the meantime President Nixon took action to sustain the national ocean program. In a policy statement made on October 30, 1968, he emphasized the need to press forward with the national oceanographic program so that economic benefits would be forthcoming as soon as possible, and so that the United States did not find itself in second place in "underwater security". The need to involve private industry in the program was stressed, but international cooperation was to be promoted only when it was in the "best interests" of the United States.[13] Taken together with the U.S. draft principles, and the abstention on the international machinery study vote, this statement illustrated the widening gulf between U.S. perceptions of an international regime, and the Pardo-inspired perceptions of the developing countries. The United States, like the Soviet Union, was interested in pursuing its large and growing vested interest in ocean activities with as little interference as possible. Both governments wanted to protect freedom of navigation for commercial and military traffic, and freedom for marine scientific research. In addition the Soviets were concerned over freedom for their new high seas fishing fleets, and the Americans were concerned with encouraging nodule mining in order to provide alternative sources of supply of the four metals. The drive of the two superpowers toward increased exploitation of ocean resources thus continued unabated, carrying with it, and stimulating, the newly emerged aspirations of the developing countries.

By the end of 1968 it was clear that a new seabed regime was not going to be created quickly. Differences on most key issues ran very deep, and the developed states were entrenching themselves against the rapidly growing and increasingly assertive demands of the developing countries. The establishment of a permanent SBC was in itself an admission that the project was going to take some time, and this in turn meant that the negotiations would have to take place in an environment in which the key data in terms of technology, economics, and vested interests would be changing and evolving continuously. It was also becoming clear that the seabed debate, both on its own merits, and in conjunction with the rapid increase in pressure from all ocean uses, was inexorably reopening the law of the sea debate as a whole.

THE GREAT SEABED DEBATE: 1966-70

Between 1966 and 1970, and largely in the United States, a great public debate took place on the issue of deep sea mining and a regime for the seabed beyond the limits of national jurisdiction. To a considerable extent this debate was a product of the fact that interest in mining manganese nodules was much more advanced in the United States than elsewhere, but it was also a byproduct of the awareness and capability created by the national ocean program started earlier in the 1960s. The debate took place mostly in academic circles, and was dominated by international lawyers, oceanographers, and economists. Academic institutions like the Law of the Sea Institute at the University of Rhode Island played a major role in encouraging and facilitating the debate, and through such forums government and industry people and representatives of various other interest groups were also drawn in.

In most respects this debate was conceptually ahead of the parallel debate going on in the United Nations. It had started earlier, was less constrained by political considerations, and had the advantage of taking place largely among people who were already experts in the field. In addition, since most of the information pertinent to the debate was being generated in the United States, it tended to influence the domestic debate before filtering through to the international one. Indeed, the products of the great debate in the form of academic articles and conference reports constituted one of the principal inputs to the Seabed Committee, both directly and indirectly as source material for the many reports of the Secretary-General on the subject.

The debate was triggered by the impending creation of an active deep sea nodule mining industry. By 1966 at least three American companies–Deepsea Ventures, Inc., Kennecott Copper Corp., and Ocean Resources Inc.–were actively prospecting for nodules and working on technology for mining and processing them. (See Chapter 3.) In 1968 Japanese companies conducted the first tests of a continuous line bucket (CLB) mining system, and in 1969 Howard Hughes' Summa Corporation joined the race by beginning exploration and equipment tests, while West German companies began commercial research on nodules. By 1970, Deepsea Ventures had tested an airlift dredge at depths up to 2,400 feet; Ocean Resources had conducted a second test of the Japanese CLB system for a large consortium of companies from several countries; International Nickel Company (INCO) and Deepsea Ventures were both well advanced with pilot research on processing plants; and CNEXO, the French nickel company, had begun exploration cruises in the Pacific.[14] A substantial momentum towards a deep sea nodule mining industry was established during these years, and the prospect of such an industry raised the legal and political questions that gave birth to the debate.

It was known from the beginning of the debate that the nodules that were of interest as ore were nearly all located on the deep seabed beyond the outer limits of any continental margin or national claim. In particular, interest was centered on the north-central area of the western Pacific. This meant that the legal regime applicable to nodule mining was uncertain. It was possible to argue that the vagueness of the exploitation criterion in the Continental

Shelf Convention allowed the provisions of the convention to apply to such activity, but this was a contentious interpretation that opened up prospects for division of the entire ocean floor by coastal states. It was also possible to argue that the High Seas Convention covered the situation, in which case mining could be allowed as a "freedom" in the same sense as fishing. The difficulty here was that the International Law Commission had specifically noted that it was not elaborating a regime for deep sea mining,[15] and consequently there was no certainty as to whether the deep seabed was res nullius, and therefore subject to claim, or res communis, and therefore open to all comers. This uncertainty was exacerbated after 1967 by the idea of the common heritage of mankind, which, although it had no legal force or meaning, was politically very powerful.

The uncertainty of existing law on the deep seabed was interpreted within the debate as both a curse and a blessing. Those in favor of the commercial industry were concerned that legal uncertainty would stifle development. They reasoned that the industry must be guaranteed security of investment before private companies would risk the large sums necessary for developing the technology and establishing mining sites. Unless the companies could be sure that their sites would be secure from claim-jumping and expropriation it would not be worthwhile their investing in nodules. In addition, since the technological problems associated with deep sea mining were unusually formidable, and since the economic profitability of the venture was uncertain even if the technological difficulties could be overcome, this school argued that the mining companies needed not just an absence of legal problems, but legal incentives as well, if they were to be encouraged to continue with an activity of great potential value to the United States. Such incentives might, for example, take the form of flag protection by the state for any mine site established by the company.

Those who saw the legal uncertainty as a blessing were largely those in favor of establishing an international regime for the area along the lines eventually suggested by Pardo in the General Assembly. For the most part they supported the idea of the seabed as common property or the common heritage of mankind, and had hopes that the wealth found there might be used to strengthen international organization or to reduce the disparity between the developed and the developing countries. They were also frequently motivated by a concern that action in support of the mining companies would lead to the ocean being partitioned among the coastal states.

If the existing law was inadequate, and most agreed that it was, the problem arose of what kind of new law to create, and how to create it. Should the new law be created by state practice based on extensions of existing law and practice, or should it be worked out in the international forum and codified into a convention? Should the new law anticipate the nascent industry or follow it? The former course would ensure control over the development of the industry, but at the risk of stunting or preventing its growth if the law did not correctly anticipate its still unknown requirements. The latter course would encourage development of the industry, but would invite conflict and leave all initiative in the hands of the companies. Law would be reduced to a passive, responding role rather than an active, guiding role.

The participants in the debate developed positions covering all points of view, from extreme nationalist ones involving huge extensions of coastal state jurisdiction, to extreme internationalist ones based on an international agency controlling all activity in the oceans. As the various positions were defined and clarified, the points of view of various vested interests also emerged. These included consumer states interested in new sources of the four nodule metals for purposes of lower prices and increased security of supply; exporter states interested in protecting the price of their products; business interests anxious to protect investments already made, eager to open up chances for profit, and convinced that increase in supply was good for all on a global scale; Marxist and other left-wing critics pointing out the potential for a new colonialism on the seabed by the technologically advanced countries; environmentalists urging the need for a global approach to protection of the vital ocean ecosystem; internationalists hoping to use seabed wealth to encourage the growth of strong international institutions; and would-be common heritage owners anxious both to prevent the looting of the common heritage by a few technologically capable states and to encourage exploitation in the hope of sharing in the benefits produced.

The most important of the alternative regime scenarios developed during the debate were as follows:

(1) National lakes. Under this scheme coastal states would extend their maritime boundaries out to the median line with their opposite neighbors. Such extension would be justified by an interpretation of the Continental Shelf Convention, and would result in the entire seabed falling under national jurisdiction. Mining activity would take place under the conditions of the convention by arrangement with the appropriate coastal state. This scenario was developed by many writers as a "scare" example but found very few advocates. Most of the countries that would gain the largest areas opposed the idea because of its inherent threat to freedom of the seas, and in general the scheme was seen as being too inequitable and too cumbersome to be a realistic alternative.

(2) Flag state regimes. This scheme had a number of variations, all based on the idea that the state in which the mining company

was registered should assume jurisdiction or sovereignty over the area of the mining company's claim. Mining would then take place under the laws and regulations of the flag state, which in turn would provide security of tenure and protection from harassment for the mining company. The flag state would claim the area under a res nullius interpretation of the high seas regime on the basis that establishment of a mine site constituted occupation of the area. The scheme was obviously oriented toward the interests of mining companies and importer states, but could be embellished by international registry and dispute settlement machinery, or by schemes for revenue sharing with developing states that accepted it. It was frequently presented as a proposal for a "transitional" regime to facilitate mining while an international regime was being worked out in the UN forum.[16]

(3) Independent regimes. This regime was offered primarily as an interpretation of existing international law regarding deep sea mining. It was based on the conclusion that it was not illegal for companies or states to undertake such mining if they cared to go out independently and take the risk of being claim-jumped by other miners. It was not generally considered a serious option because of the likelihood that aspirant miners would be deterred by insecurity of tenure.[17]

(4) Regional regimes. The idea of regional regimes was presented as a transitional mechanism to fuller international regimes, which might take too long to develop in one leap. The basic proposal was for the states of a region to extend their mining laws by agreement to the seabed area adjacent to the region. This was a substitute for an international licensing regime and did not exclude an international registry.[18]

(5) International registry regimes. These regimes were still very much oriented toward mining and importer state interests, and made only minimal concessions to the common heritage. The basic idea was for a weak international organization to fulfill essentially passive functions such as registration of claims, arbitration of disputes, and supervision of standards for environmental and safety regulations. Initiative would remain with mining companies and states, and the obligation to the common heritage would be served by contributions to some form of international fund. This option, or variations on it, found considerable support among mining and importer state interests as a compromise acceptable to them between the flag state regimes and stronger international regimes.[19]

(6) International licensing regimes. This category was the one in which the largest number of alternatives was explored. The basic idea was of an international authority that would be empowered to license exploitation of the deep seabed, but there were wide variations

on the scope of powers to be given to the authority. In weak licensing regimes the authority would be obliged to allocate licenses almost on demand, and would differ little from a registry regime. In strong licensing regimes the authority would have considerable powers of discretion over allocation of licenses, and, like a state, would retain most powers of initiative for itself. Additional proposals associated with strong licensing regimes included vesting sovereignty or jurisdiction over the seabed in the authority, giving it powers to control production in order to protect land-based producers, and making it the mechanism through which revenues derived from seabed exploitation would be distributed among the owners of the common heritage. It was in this category that proposals began to appear that envisaged the seabed authority as a large, innovative, and important new international organization.[20]

(7) International exploiting regimes. The core idea of this regime was that the international authority itself should undertake exploitation of the deep seabed. The authority would thus become the embodiment of the common heritage, and the regime would eliminate any possibility of the common heritage falling victim to the skills of the developed states. This proposal, which was very much anti-developed country in its intent, raised the problem of how the authority would be able to undertake exploitation without using the capital and technology of the developed states.

(8) International ocean regimes. These regimes sat on the most internationalist end of the spectrum, and represented a considerable extension of the seabed authority idea. The proposal was for an authority that would regulate or control all activities in ocean space, and not just those relating to the seabed. Many of the arguments in support of this type of regime were based on the interrelatedness of all aspects of ocean ecology and use, and on the consequent need for an overarching mechanism to ensure protection and efficient management of the oceans as a whole. Such proposals were a significant development in the tendency noted above for discussion of the seabed to spill over into other aspects of the law of the sea.[21]

The debate divided between national and international regimes, and it became clear after Pardo's proposal that only an international regime of some sort would be likely to find general approval among the nations of the world. No state had expressed interest in pursuing the national lakes solution, and given the aroused interest of the developing countries, a flag state regime seemed likely to invite excessive controversy and conflict. As a result, the debate concentrated mostly on the four variations of the international regime, although this should by no means be taken as an indication that the flag state option had been dismissed. Its advocates simply changed their tactics and offered their scheme not as an alternative to international regimes, but rather as an interim regime, or as an option if an international regime proved impossible to negotiate.

The general acceptance of an international regime of some sort as the object of the exercise had a profound impact on the debate because of the number and type of questions it automatically raised. As soon as international machinery was assumed, the questions arose of how it was to be organized, financed, and staffed; what its exact mandate would be, and what powers it would be given to fulfill it; and what its relationship with existing organizations would be. Since the regime and machinery would close off a large area of the planet to national claims it was more necessary than with more functional international organizations that it have nearly universal support. This raised the question of how it could be designed to attract such support given the variety of conflicting expectations (equity, economic efficiency, environmental protection, and so on) surrounding it. On a more specific level questions arose as to how licenses would be allocated if a licensing regime were chosen; how, and in what form, benefits from seabed exploitation would be distributed to the owners of the common heritage; and how the authority would enforce its rules and regulations.

The most important question raised by the prospect of an international regime, however, was the extent of the area to which it would apply. This question was crucial for two reasons. First was the political reason that the boundary of the regime would define the precise outer limits of national jurisdiction on the seabed. Second was the economic reason that the resources falling under the international regime varied drastically according to where the boundary was drawn. The main factor here was whether or not the regime would derive revenue from continental shelf oil, and until this was decided it was difficult to design an appropriate machinery. These two aspects, the economic and the political, created a dilemma that was to play a major role in paralyzing the SBC. The problem was that states would not take the political and economic risks of agreeing to a fixed boundary for the area until they knew the nature of the regime that was to govern it. Because of the economic factors involved, however, it was exceedingly difficult to define the nature of the regime until the area to which it was to apply, and therefore the nature of its resources, was known.

The boundary question was also important because of the impetus it provided to reopen issues of coastal state jurisdiction. Since fixing a boundary for the area was in effect the same as fixing a boundary for the continental shelf jurisdiction of coastal states, the process automatically invalidated the compromise reached in 1958 by leaving the definition flexible. This led to the exhuming of all the fixed limit alternatives that had been buried under the exploitation criterion, and consequently to the renewal of the narrow-versus wide-limits dispute. The dispute had lost none of its vigor despite the fact that its context was markedly different from what it had been in the 1950s. By the late 1960s there were vested interests on both rather than one side of the boundary.

Although it was essential to settle the question of coastal state limits in order to construct an international regime, the raising of the question turned out to be a major blow to hopes for an international regime. To a considerable extent it facilitated the turn away from the extreme difficulties and uncertainties of the international regime debate that occurred after 1969. By bringing attention to the much more readily understood political and economic realities of the continental shelf, it encouraged a shift toward debate on a coastal zone. This shift took place gradually, through such mechanisms as proposals for intermediate zones where the jurisdiction of the coastal state would blend into that of the international authority, but its effect was to drain the international regime debate of much of its force and urgency and to leave it increasingly in the shadows of the main renegotiation of the law of the sea.

Another problem afflicting the international regime debate was the uncertainty about the economic value of the resources of the deep seabed. This uncertainty was in addition to that created by the unknown size of the area, and related primarily to evaluations of the prospects from a nodule mining industry. The size of the deposits was not questioned. As in the traditional quandary about the amount of gold contained in a cubic mile of seawater, the question was: could it be recovered economically? Technology for recovering the nodules constituted part of the uncertainty, but given the rate of advance in this area it seemed likely that bulk production systems would be available within a few years. The main uncertainty lay with the economics of the operation, and particularly in the likely amount of surplus revenue it would generate. On this front almost nothing was known, and argument had to be based largely on conjecture.

The interest of the companies in nodules of itself indicated that the prospects for profit were at least reasonable, and probably considerable. The difficulty was that since production costs would not be known accurately until the industry was under way, it was impossible to predict how much revenue the authority might reasonably expect to receive. Optimists and pessimists on this subject were among the institutions of the debate[22] and their predictions were tied closely to the positions on a comprehensive or a limited international regime and machinery. If the pessimists were correct, then many of the more ambitious schemes would scarcely be able to pay for themselves, let alone generate revenue for the common heritage. If the optimists were correct, however, then a weak regime might enable the companies and the developed states to take more than their share of the common heritage. There was no way of resolving the dispute until hard economic data became available from practice; in the meantime the debate suffered from a plethora of conflicting speculations.

This uncertainty surrounding the central economic objective of the deep seabed interest had a number of important impacts on the debate. First, it made the debate exceedingly abstract, theoretical, and conditional in nature, and therefore frustrating and prone to conflict in practice; this had the effect of slowing it down and sapping its momentum. Second, it opened the international debate to an undue amount of influence from factors related to the politics of U.S. economics. U.S. mining companies were a prime source of information on nodule economics, and especially in the early stages of the debate their optimism, which was logical in view of such needs as attracting capital and government support, tended to engender excessive optimism among many international regime planners. Third, it encouraged the politicization of information about nodule mining that increasingly became a feature of the debate. Since all information was speculative, it became easy to associate information that supported a particular view with the view itself (for example, pessimism with desire for a weak regime). Thus all information became politically suspect as having been manufactured to fit a given point of view, and consequently new information ceased to have a rationalizing effect on the debate.

The great debate was pursued with vigor all through the late 1960s. Various components of it continued well into the 1970s, but after 1970 it was superceded by the broader debate on the whole law of the sea. To a certain extent the debate fell victim to its own internal difficulties. It was unable by its nature to reach any conclusion, and many of its motives were made redundant by the ongoing work of the SBC.

Despite the lack of drama in its own outcome, the debate played an important role in the development of seabed politics. Although it was basically part of the interest articulation/aggregation process relating to policy-making in the United States, it also had a substantial effect on the negotiations in the SBC and the General Assembly. Within the context of the debate options were explored that could not have been explored in the international forum. Problems attending various options were identified and discussed before they arose in the SBC, and a considerable effort was made to ensure that the fruits of the debate were made available as widely as possible. The debate was an important component of the educational part of the SBC's work, and in general it added depth, scope, and information to the international negotiations. It provided the major source of information, both hard and speculative, about manganese nodules, but because of the political elements it introduced, this function may have proved as much a disfavor as a favor to the work of the SBC.

NOTES

1. The two resolutions eventually resulted in UN Docs. E/4449 and Adds. 1 and 2 (February 1968), and E/4487 (April 1968).

2. UN Doc. A/6695 XXII, August 18, 1967.

3. UN Monthly Chronicle 5, no. 1 (January 1968), p. 29; and interviews with delegates.

4. Interviews with delegates.

5. UN Doc. A/6695 XXII, op. cit.

6. Ibid.

7. UN Monthly Chronicle 5, no. 1 (January 1968), pp. 29-32; and Evan Luard, The Control of the Sea-bed (London: Heinemann, 1974), pp. 83-90.

8. GAOR (General Assembly, Official Records) XXIII, A/7230, September 1968 (Report of the AHSBC), pp. 17-19.

9. On the AHSBC's work generally during 1968, see ibid. and AHSBC summary records A/AC.135/SR.1-26, A/AC.135/WG.1/SR.1-14, and A/AC.135/WG.2/SR.1-15.

10. A/AC.138/7 and Corrs. 1-4 (March 1969), pp. 15-25.

11. UN Monthly Chronicle 5, no. 1 (January 1968), pp. 29-30; and William E. Butler, The Soviet Union and the Law of the Sea (Baltimore: Johns Hopkins Press, 1971), p. 156.

12. Butler, ibid., pp. 144-45; and Butler, "Some Recent Developments in Soviet Maritime Law," in Lewis M. Alexander (ed.), The Law of the Sea: The United Nations and Ocean Management (Kingston: University of Rhode Island Press, 1971), p. 382.

13. Text reprinted in Norman J. Padelford, Public Policy for the Seas (Cambridge: MIT Press, 1970), pp. 336-38.

14. U.S. Senate, Mineral Resources of the Deep Seabed, Hearings before the Subcommittee on Minerals, Materials and Fuels of the Committee on Interior and Insular Affairs, 93rd Congress, first session on S.1134, May 17, June 14, 15, 18, and 19, 1973, Washington, D.C.: USGPO, 1973, pp. 202-04.

15. ILC Report for 1956, GAOR XI Supp. 9, A/3159, p. 24.

16. See, for example, Haight, Laylin, and Young in Shigeru Oda, The International Law of the Ocean Development (Leiden: Sijthoff, 1972), pp. 299-305; and Northcutt Ely in Natural Resources Lawyer 81 (June 1968) or summarized in Erin B. Jones, Law of the

Sea: Oceanic Resources (Dallas: Southern Methodist University Press, 1972), pp. 126-27.

17. See, for example, Louis Henkin, "Changing Law for the Changing Seas," in Edward A. Gullion (ed.), Uses of the Seas (Englewood Cliffs, N.J.: Prentice-Hall, 1968), pp. 81-82; and Edward D. Brown, The Legal Regime of Hydrospace (London: Stevens, 1971), pp. 82-103.

18. See, for example, L. F. E. Goldie as summarized in Jones, op. cit., pp. 124-25.

19. See, for example, proposals by the American Branch of the ILA, the ILA, and the American Bar Association in Oda, op. cit., pp. 251-71; and J. L. Mero, "A Legal Regime for Deep Sea Mining," San Diego Law Review 7, no. 3 (July 1970).

20. See, for example, proposals in Reports 19, 21, and 23 of Building Peace: Reports of the Commission to Study the Organization of Peace 1939-1972 (Metuchen, N.J.: Scarecrow Press, 1973); and Aaron L. Danzig in Oda, op. cit., pp. 244-51.

21. See, for example, E. M. Borgese in Oda, op. cit., pp. 280-99; Francis Christy, Jr. in Natural Resources Lawyer (June 1968); and Claiborne Pell in Oda, op. cit., pp. 272-80. For a general discussion of the alternative regime options, see also L. F. E. Goldie, "The Contents of Davy Jones' Locker—A Proposed Regime for the Seabed and Subsoil," Rutgers Law Review 22, no. 1 (Fall 1967), pp. 1-66; and several of the articles in Natural Resources Lawyer 1, no. 2 (June 1968).

22. Brown, op. cit., pp. 103-07.

CHAPTER

5

FROM SEABED TO OCEAN POLITICS, 1969-70

From 1969 to 1970, the SBC remained stable with the form and mandate given to it by resolutions 2467A and 2467C. It decided to carry on with the structure created by its predecessor, and so reconstituted the legal, economic, and technical working groups into subcommittees. Aside from the changes in mandate and membership resulting from the 1968 resolutions, the Seabed Committee was very much a direct continuation of the AHSBC. It continued both the method of working by consensus, and the whole framework of discussion laid down by the AHSBC; its work followed the three themes of acquiring information, negotiating a declaration of principles, and defining areas of disagreement. A key trend during these years was the steady swing toward a broader discussion of all law of the sea issues, and a corresponding drop in the priority given to the more specialized matter of the international seabed regime and machinery.

SEABED POLITICS IN 1969

The SBC held six weeks of meetings during 1969, and after settling in to its new form, quickly picked up the work begun by the AHSBC. The major source of information for the committee continued to be the reports and studies submitted by the Secretary-General and the specialized agencies, supplemented by the speeches of the more technically informed delegations. The Secretariat produced a major study on the question of international machinery (A/AC.138/12) and shorter studies on economic considerations affecting exploitation (138/6), marine pollution arising from exploitation (138/13), and national legislation (138/9, a supplement to 135/11). In addition, the IOC submitted a report on a long-range program of ocean research and exploration (138/10 and 14). The study on international machinery was produced in response to Resolution 2467C of the previous December, but was not finished until June 1969, more than halfway through the SBC's sessions for the year. It covered possible functions, powers, and structures of an international seabed authority, and provided an important focus for the year's debate. Even with the previous year's experience and the new round of studies, the committee's work in 1969 was still largely dominated and shaped by the need for more information and the necessity to educate those delegations not familiar with marine affairs.

Progress on a declaration of principles was slower than many had hoped. The Legal Subcommittee devoted almost all its attention to this item, but was unable to make much advance on the position established by the AHSBC. No new formal proposals were submitted, and in its report, the subcommittee had to content itself with a tentative list of common denominators.[1] According to this list agreement seemed to exist on the following:

(1) That an area did exist beyond the limits of national jurisdiction;

(2) That the area should not be subject to national appropriation by any means;

(3) That principles and norms of international law existed that applied to the area;

(4) That the area should be reserved for peaceful purposes;

(5) That a regime for the area should be established, and that it should ensure the use of the resources of the area for the benefit of mankind, taking into account the interests and needs of the developing countries;

(6) That there should be freedom of scientific research in the area; and

(7) That use of the area should be subject to safeguards regarding pollution, living resources, and the other uses of the high seas. Beyond this very general level, however, there was little agreement on definitions. Disputes existed as to which laws applied to the area, how peaceful purposes should be defined, what constituted national appropriation and whether there should be a freeze on further seabed claims and mining activities, what kind of regime should be devised and whether it should cover the area itself or only the resources, and how scientific research could be distinguished from commercial research. As in the previous year, the debate on a declaration of principles proved more productive in defining areas of disagreement than in finding grounds for compromise.

The dispute over the common heritage principle remained a feature of the debate in the SBC, and some progress was made toward closing the gap between developed and developing countries on the subject. The Soviet Union continued to lead opposition to the principle, while Kenya, Malta, and Tanzania were foremost in its defense. So long as there was no agreement on what the developing countries saw as a fundamental principle, an emotional block to further progress existed. The developed states appeared increasingly to recognize the special quality of this principle during 1969, and many of them began to reconcile their difficulties with it by associating it with the idea of using the resources of the area for the benefit of mankind, taking into account the needs and interests of the developing countries. The process was a slow one, and the common heritage principle was still a contentious issue at the end of the 1969 SBC sessions.

Disagreement over the question of establishing a boundary for the area remained very strong during 1969. The Soviet Union joined the Western states in favoring such a boundary, and Malta incurred the displeasure of the United States by suggesting a 200-meter/200-mile depth/distance formula as a possible solution. On the other side, Chile, Argentina, and Brazil attacked the reference to a 12-mile limit contained in the draft seabed demilitarization treaty considered by the committee in November. The committee remained completely stalemated on how to define a boundary for the area, but did manage to agree on a recommendation that the question be included on its agenda. It also discovered the dilemmas inherent in defining the area before the machinery or vice versa, but made little progress toward resolving them.

Although the SBC itself remained unable to settle the boundary issue, a decision by the International Court of Justice in February 1969 strongly influenced the debate. The court was considering the North Sea continental shelf case submitted to it in 1967 by Denmark, West Germany, and the Netherlands. The problem was the failure of the three states to agree on the delimitation of their seabed boundaries. In its decision on the case the court stated that:

> the rights of the coastal state in respect of the area of continental shelf that constitutes a natural prolongation of its land territory into and under the sea exist ipso facto and ab initio, by virtue of its sovereignty over the land. . . . There is here an inherent right.
>
> What confers the ipso jure title which international law attributes to the coastal state in respect of its continental shelf, is the fact that the submarine areas concerned may be deemed to be actually part of the territory over

> which the coastal state already has dominion,—in the sense that, although covered with water, they are a prolongation or continuation of that territory, an extension of it under the sea.[2]

This interpretation, and in particular the idea of "natural prolongation" of territory already under sovereign control, gave a great boost to those states favoring or claiming control out to the edge of the continental margin.

The question of peaceful use of the area became less important as a focus of conflict during 1969. While all states agreed with the principle of peaceful use, it was clear that no detailed agreement on the implementation of such a principle could be achieved. The burden of this issue was increasingly removed from the SBC by treaty negotiations, in the Eighteen Nation Disarmament Committee (ENDC) and the General Assembly, on the prohibition of the emplacement of nuclear weapons and other weapons of mass destruction on the seabed. The SBC met in November to consider the draft treaty presented to the conference of the Committee on Disarmament by the United States and the Soviet Union, and the Soviet delegation continued to emphasize the issue, but in the Seabed Committee as a whole there was a tendency simply to accept the principle without making any attempt to elaborate its implications.[3]

By contrast, the dispute over international machinery expanded to become the most visible conflict within the committee. Stimulated by the General Assembly debate on Resolution 2467C and by the subsequent report of the Secretary-General, the dispute over international machinery overshadowed the disagreements on the common heritage and the boundary issue. By outlining alternative structures, powers, and functions for such machinery, the Secretary-General's study provided a needed framework for the debate, and resulted in clearer positions and alignments on the subject than had been possible the previous year.

Only the Soviet Union continued to oppose the idea of any international machinery, and so during 1969 the question became not so much whether to have machinery, but rather what type of machinery to have. The developing countries expressed themselves firmly in favor of having machinery, and Brazil and Chile led those insisting that the machinery must be linked inseparably with the declaration of principles. (This was the developing countries' equivalent to the insistence of the developed states that the area must be given a precise boundary.) India and Kuwait criticized the Secretary-General's report for its emphasis on the weaker forms of machinery, and along with Sri Lanka expressed interest in more information on machinery that would undertake exploitation. The United States argued strongly against exploiting machinery on the grounds that it would be economically inefficient and would absorb all the revenues generated from seabed exploitation. Other difficulties arose over whether the regime and machinery should apply to the area itself, as India and Mauritania insisted, or just to the resources of the area.

The machinery dispute continued to reflect a division in the SBC along developed/developing lines. In general, the developed states were less enthusiastic about the prospect of machinery, and favored weaker forms of it, while the developing countries were very enthusiastic about the prospect of machinery, and wanted it to be strong and comprehensive. This division accurately reflected interests in the matter. The developed states needed little in the way of machinery to enhance the option to exploit that command of technology gave them, while the developing countries had no option to participate in exploitation outside of an exploiting machinery, and were concerned that weak machinery would exclude them from all but symbolic control over exploitation and the distribution of benefits.

Progress in the debate, and an increasing level of understanding of the issues involved, resulted in the emergence of new problems. Two of these were becoming visible in 1969: the idea of transferring marine technology from the developed states to the developing countries, and the possibility of seabed mineral production causing economic hardship for developing country exporters of the four nodule metals. The transfer of technology was raised by Liberia and Nigeria, but while it attracted a certain amount of sympathy from the developing countries, it did not result in much discussion of the formidable problems involved. The problem of existing exporters was first raised by Argentina in 1968 and later echoed by Peru (a large copper exporter) in the SBC. It took on a new dimension when viewed in the light of a possible exploiting machinery, since this raised the possibility of an international organization helping some developing countries at the expense of others by competing in their export markets. This in turn raised the issue of production and marketing controls by the seabed authority, a proposal that elicited strong disapproval from the United States and other developed states.

The results of the SBC's year of work were thus again inconclusive. No draft declaration of principles could yet be recommended to the General Assembly, the debate on international machinery was still in its early stages, and little progress had been made on settling the other major disputes confronting the delegates. The Soviet Union was still bearing the brunt of the opposition to the developing countries, with the United States in close second place, and the Latin American group continued to be in the forefront of the developing countries' side, along with Kenya, Tanzania, India, Sri

Lanka, and Kuwait. There were accusations on both sides that states were taking hard lines in order to stall the work of the committee. The Latin American states were seen as stalling in order to gain more converts to their wide coastal zone position, while the developed states, particularly the United States, were seen as stalling in order to build up momentum in their own deep sea mining projects. Given the nature of the debate, it was difficult to confirm or deny such accusations. Adherence to a strong position might result from a genuine belief in its validity, or it might simply be a calculated delaying tactic. In either case the consensus method provided little incentive for states to abandon their positions at such an early stage of the negotiations.

The education functions of the SBC, both in terms of its own members and in terms of its reporting obligations to the General Assembly, were assuming a permanent position in the committee's work. Not only did influxes of new members have to be socialized into the committee's background and procedures, but the continuous flow of new information on the environment, resources, and economic potential of the deep seabed had to be reviewed, absorbed, and integrated into the debate. One result of this steady input of new information was a diminution of the initial optimism about the prospects of great wealth from the seabed. Wealth from nodules increasingly looked like an uncertain and distant prospect at best, subject to the solution of numerous technical difficulties in mining and processing and under the shadow of rising concern from developing mineral exporting states. Wealth from oil, which was clearly the most promising seabed resource, was subject to sacrifices by coastal states in terms of accepting relatively narrow limits of coastal jurisdiction. Since the whole question of boundaries was effectively frozen by the deep disagreements involved, the chances here did not encourage optimism either, and the ICJ judgment on the North Sea case dampened hopes even further.

This decline in the expectations of wealth for all from the common heritage had the effect of removing some of the economic urgency from the committee's work and emphasizing instead the more political aspects of the matter. The emergence of concern over the transfer of technology and over the need for production and marketing controls was symptomatic of a gradual shift in the objectives of the developing countries away from simply assuring themselves a share of the benefits toward a concern that seabed mining should be prevented from damaging their interests as mineral exporters. Tied to this was a general concern that the developed states should not be allowed to reap any undue advantage from the common heritage simply because they possessed the ability to exploit it.

Malta raised again the question of convening a conference, and suggested that its agenda be limited to revising the 1958 Convention on the Continental Shelf so as to produce a fixed depth/distance boundary, and to formulating legal norms for the exploration and exploitation of the international seabed area. This question did not occupy much time in the SBC, but was a major item in the General Assembly deliberations following.[4]

When the seabed debate moved out of the SBC and into the General Assembly in the autumn of 1969, it carried with it even more controversy than it had in the previous year. Once again released from the gentlemanly shackles of the consensus procedure, the disagreements and conflicts between the developed and the developing countries surfaced into public view with a vengeance.

The Assembly passed two uncontroversial resolutions on December 13, 1969: Resolution 2560 supporting a long-range program of ocean exploration and UN involvement in it, and Resolution 2566 requesting a study on marine pollution. The main focus of its attention, however, was on the four resolutions that eventually emerged as 2574A-D.

Resolution 2574A grew out of the proposal by Malta in the SBC (138/11) that a new conference on the seabed aspects of the law of the sea should be convened. The Maltese delegation raised the matter again in the First Committee, but found itself outmaneuvered by a large group of developing countries. Led by Brazil, Jamaica, and Trinidad and Tobago, this group wanted a new conference covering all aspects of the law of the sea, not just the continental shelf boundary and international seabed area agenda favored by Malta and most of the Western states. The developing countries argued that the essential unity of ocean issues made it impossible to deal with the seabed in isolation from other aspects of the law of the sea, and they used their voting majority to revise Malta's proposal along these lines.

Their draft, sponsored by 17 developing countries, requested the Secretary-General to ascertain the views of states on an early convening of a comprehensive conference on the law of the sea. It noted the inadequacies of the flexible limit embodied in the Continental Shelf Convention, but clearly favored the Latin American view that establishment of an international regime for the area should take place before the area itself was defined. The preamble to the request exhibited notable restraint in that it did not mention the common heritage by name, but it did include a statement noting the "urgent necessity of preserving this area from encroachment, or appropriation by any State, inconsistent with the common interest of mankind."

The debate surrounding this proposal was heated, with Malta denouncing the revision of its proposal as "a political maneuvre".[5] When it came before the General Assembly as Resolution 2574A on December 15, the proposal passed by 65 to 12 to 30, making it the second important resolution on the law of the sea to be carried on the strength of the developing countries' majority.[6] Like Resolution 2467C, 2574A was a successful exercise of vote power over technological, economic, military, and other forms of power because its objectives were confined within the purview of the General Assembly. Its passage significantly altered the whole context of the SBC's work by making explicit the extent to which the seabed debate had reopened the whole law of the sea debate. The committee would now have to conduct its deliberations against the background of a wider law of the sea debate that would inevitably divert attention away from the international area toward traditional law of the sea preoccupation with the rights of coastal states. Whether this development would facilitate or disrupt the committee's work was impossible to predict. On the one hand it might facilitate it by providing a greater opportunity to make tradeoffs among positions on which disagreement was severe—the so-called package deal approach. On the other hand, it might introduce so many factors into the debate that it would become interminably bogged down in the complexities of such a broad subject. On this issue as on the international machinery and boundary questions, accusations were made on both sides that certain states (usually the United States or Latin American states) were deliberately trying to stall the negotiations.

In either case it was clear that the developing countries were determined to undertake a thorough reexamination of the law of the sea. A great many of them had not participated in the 1958 conference that had codified the existing law of the sea, and many of those that had participated were not satisfied with the outcome. There was a general feeling among the developing countries, especially the coastal states, that the existing law of the sea was biased toward the maritime states; since very few of the developing countries had significant maritime capabilities, the existing law was seen as working against them.

Resolution 2574B simply renewed the SBC's mandate and urged it to complete its work on a draft declaration of principles in time for the next year's session of the General Assembly. This was not controversial, and passed 109 to 0 to 1.

Resolution 2574C was an extension of Resolution 2467C, which had proved so controversial the previous year. It was a request for a further study by the Secretary-General on the question of international machinery, this time giving particular emphasis to a machinery with the power "to regulate, coordinate, supervise and control all activities relating to the exploration and exploitation of resources." Although not explicitly stated, this amounted to a request for a study of exploiting machinery as favored by many of the developing countries, and was in some senses a criticism of the previous report (138/12) that had concentrated on weaker types of machinery.

The resolution grew out of the sharp debate on international machinery that had been carried over from the SBC to the First Committee of the General Assembly. The Soviet Union and its allies continued to oppose the whole question as premature, while the United States, Britain, France, Japan and others resisted the demands of the great majority of developing countries—led by Brazil, India, and Kenya—for a machinery with power to exploit. Forty-eight developing countries sponsored 2574C (18 African, 11 Asian, 13 Latin American, five Arab, and Yugoslavia); except for Turkey, they were all members of the Group of 77. The resolution passed by 100 to 0 to 11, and the contrast between this vote and the 85 to 9 to 25 result on 2467C indicated the extent to which the idea of machinery had gained acceptance despite the severe disagreements on its most desirable form.[7]

Resolution 2574D, known as "the moratorium resolution", was easily the most controversial of the four, and reflected a substantial deepening of the rift between developing and developed countries. Its 12 developing country sponsors unabashedly stated their conviction that it was essential for exploitation of the area to "be carried out under an international regime, including appropriate international machinery." The resolution went on to declare

> that, pending the establishment of the international regime,
>
> (a) States and persons, physical or juridical, are bound to refrain from all activities of exploitation of the resources of the area of the sea-bed and ocean floor, and the subsoil thereof, beyond the limits of national jurisdiction;
>
> (b) No claim to any part of that area or its resources shall be recognized.

At first glance, the text appeared to be a strongly worded but relatively straightforward moratorium on both claims and exploitation activities. After a deeper reading, however, the realization that the area under discussion had not been defined rendered the idea of a moratorium virtually meaningless in practical terms. It was impossible by definition for any national claim to violate the moratorium, and therefore the resolution could be interpreted as encouraging either flag state claims to protect mining activity, or extravagant interpretations of the Continental Shelf Convention. The dilemma in the text was sufficiently obvious to make oversight on the part of the drafters an unlikely cause. A better explanation seemed to lie in the objectives and capabilities of the resolution's supporters.

Of the 12 sponsors of 2574D, three (Chile, Ecuador, and Peru) were long-time exponents of large national claims, and a fourth (Brazil) was on the brink of making its own 200-mile claim. It is hardly credible that in sponsoring this resolution these states were abandoning their long and increasingly successful struggle to encourage more states to make wide coastal zone claims. Chile, Ecuador, and Peru had opposed a claims moratorium when the idea was raised during the 1968 AHSBC discussion, and nothing in their subsequent behavior suggested that they had altered their policies. The deliberately ambiguous wording of Resolution 2574D thus served their purposes admirably.

But while the loophole in the moratorium facilitated coastal state claims to an adjacent zone or an extended territorial sea, it was not so conducive to mid-ocean claims of the type that would be necessary to facilitate nodule mining. Such claims would clearly fall within the area that all states agreed was beyond the limits of national jurisdiction in the conventional sense (of adjacency), and therefore they would be much harder to justify under the terms of the moratorium. Resolution 2574D, then, was clearly aimed at restricting nodule exploitation by a handful of developed states, and not at restricting claims to coastal jurisdiction by the developing countries.

Support for this interpretation can be found in testimony by Professor L. F. E. Goldie, who noted that

> one of that Resolution's promoters has recently conceded, in a statement for which no ascription can be given, that the Moratorium Resolution was intentionally and specifically directed against the private enterprise interests of the United States and other private enterprises of developed countries. It was framed to have a "chilling" effect on investors.[8]

In this sense, the moratorium resolution marked an important departure in the struggle between the developed and the developing countries for control over the deep seabed. It represented the first attempt by the developing countries to bring their voting power to bear on a situation wholly outside the purview of the General Assembly. The resolution was not legally binding on states, and indeed could not be given the ambiguities in its drafting. What it amounted to was a more subtle effort to undermine the technological advantage of the developed states by increasing the uncertainty surrounding the legal environment in which nodule mining would have to take place. Legal uncertainty was a noted feature in the internal debate within the United States, and therefore it was a reasonable calculation to assume that exacerbating it would discourage investment in the nodule industry.

Voting power was the only weapon available to the developing countries with which to combat the steadily increasing economic and technological advantage of the developed states. It was a way of buying time in which to negotiate before the whole negotiating process was overtaken by a fait accompli; in the context of the strong move toward more comprehensive negotiations represented by Resolution 2574A, such time might prove crucial to success. It was also a way of making a strong statement of determination on the deep seabed issue. At best, the moratorium resolution was a successful piece of psychological warfare that had some effect in slowing down the nodule mining industry. At worst, it provided a useful propaganda club with which to beat the developed states.

In either case, however, the advantage was not won without costs. The resolution heightened the sense of conflict in the negotiations both by the pointed character of its attack and by the sharp cleavage between the developed and developing states that it precipitated. The final vote on it was 62 to 28 to 28. All but two of the supporting votes (Sweden and Finland) were cast by developing countries, and all but one of the opposing votes (Ghana) were cast by developed states. In terms of national attributes, offshore oil, geographical characteristics, and mineral import and export interests, states were heavily divided along developed/developing lines in the vote. This sense of conflict inevitably made compromise, and therefore negotiation, more difficult, and helped to cast the negotiations in a confrontation-oriented mold.

A second cost was the split in the developing countries' ranks represented by 24 of the 28 abstention votes. A substantial number of developing countries, including important states like Indonesia, Iran, Nigeria, Egypt, and Saudi Arabia, did not support the radical tactics favored by the Latin American states. A third cost, related to the first, was the devaluation of General Assembly resolutions by making them the focus of conflict rather than consensus. This cost, however, had to be seen in the light of the fact that the developing countries commanded few other powers to bring pressure to bear on the deep seabed question.

Resolutions 2574A-D were a fitting cap to a bleak year for negotiations on the deep seabed. Declining optimism, scant progress, and proliferating difficulties in the SBC were followed by confrontation and deepening conflict in the General Assembly. As a result of Resolution 2574A, the developed states were being dragged toward a comprehensive conference on the law of the sea that many of them did not want, and as a result of Resolution 2574D they faced the danger of becoming permanent victims of an aggressive voting majority not in sympathy with their interests. Although the developed states had the power to ignore the moratorium resolution, they were politically on the defensive in both the SBC and the General Assembly. By adroit use of their voting power, the developing countries continued to retain the initiative in seabed politics, and to move the negotiations in a direction desirable to them.

SEABED POLITICS IN 1970

During 1970 the Seabed Committee held seven weeks of meetings. It was now working in the context of the move toward a comprehensive law of the sea conference established by Resolution 2574A, and against the background of confrontation created by Resolution 2574D. The greater interest generated by the shift toward law of the sea questions was indicated by the presence of observers from two dozen states at the committee's meetings. The committee was also under some pressure of time from Resolution 2574B, which had urged it to complete a draft declaration of principles during its 1970 meetings. The stage appeared set for a repeat performance of the previous year in which the three themes of the committee's work—information and education, the declaration of principles, and the definition of areas of disagreement—would be elaborated at length, but without significant progress toward agreement. New developments, mostly arising outside the SBC, however, served to create movement and prevent a complete stalemate.

The Secretariat continued its efforts to keep pace with demands for information coming from the SBC. During the 1970 sessions, reports were submitted on government measures pertaining to continental shelf mineral resources (138/21), possible methods for sharing benefits derived from exploitation of the international seabed area (138/24), and international machinery (138/23). The report on international machinery, in response to Resolution 2574C, gave detailed consideration to more comprehensive types of machinery, including powers to exploit directly and to take action to prevent commodity price fluctuations resulting from seabed mining. The report on benefit sharing reflected the growing uncertainty and lack of optimism surrounding the stakes of the seabed debate:

> Unless the volume of seabed mineral production reaches proportions considerably higher than now anticipated it would hardly appear practical to attempt to distribute the residual proceedings accruing to the international community directly to the countries of the world, or even to the developing countries alone. . . . Definite arrangements for the sharing of proceeds may have to wait until the various factors determining the extent of economic development of mineral resources of the seabed . . . can be appraised with more precision.[9]

Progress on the declaration of principles was again slower than many had hoped, and at the end of the 1970 sessions the SBC was still not in a position to recommend a draft to the General Assembly. The Legal Subcommittee devoted its entire effort to the declaration in a concerted attempt to improve on the synthesis outlined in the SBC's 1969 report. Two draft declarations were submitted to the subcommittee—one by Norway stressing, among other things, the rights of coastal states in relation to the regime, and one by a group of 15 developing countries, including many of the most active in the SBC, stressing strong machinery, benefits for the developing countries, protection from price fluctuations, and the other familiar elements of the developing countries' position. Informal drafting groups met all through June, July, and August in the hope of formulating acceptable drafts, but there was opposition to this method of procedure (on the grounds of secrecy and lack of representativeness), and although some progress was made, no agreed declaration was forthcoming. Over the protest of Tanzania, and after much mutual recrimination between the Group of 77 and the United States over who was to blame for the impasse, the SBC decided to resume informal negotiations on the declaration of principles at the General Assembly session.[10]

By the end of the 1970 sessions, however, the difficulties with the declaration of principles were no longer at the center of the major disagreements within the SBC. Agreement was close on most of the general formulations required for a statement of principles, and consequently most disputes occurred within the framework of an overarching principle.

There was still some dispute over the common heritage, with the Soviet Union and its allies, along with Italy, Belgium, and Malaysia, opposing its inclusion as a principle, but not necessarily as a guideline. The United States did not accept the common heritage as a principle, but did not oppose it either, and other key developed states like Britain and Canada also came around to a non-opposition stance. The Argentine delegate pointed to the continuing dispute over the common heritage principle as the main cause of the committee's failure to produce a draft declaration, but it was obvious by the end of the 1970 sessions that this issue was nearly settled, and that the major difficulties lay in the more detailed disputes over how the common heritage should be put into operation.

The question of defining a boundary for the area was still very much alive, and as a result of developments outside the SBC became one of the major items in the 1970 debates. The Latin American coastal states produced two major declarations on this subject: the Declaration of Montevideo on May 8, and the Declaration of Lima on August 8. The Declaration of Montevideo resulted from a meeting of the nine Latin American states that had claimed 200-mile zones or territorial seas up to 1970 (Argentina, Brazil, Chile, Ecuador, El Salvador, Nicaragua, Panama, Peru, and Uruguay). The Declaration of Lima was the first fruit of the diplomatic offensive launched by the 200-mile states, and was supported by the nine Montevideo states plus Mexico, the Dominican Republic, Colombia, Guatemala, and Honduras.

The two declarations were similar in content and wording, amounting essentially to a claim that coastal states had the right to determine the limits of their own sovereignty or jurisdiction over both the seabed and the water column. This right was seen as stemming, among other things, from the geographical, economic and social links between the sea, the land, and its inhabitants; from the need of the coastal state to protect itself from overexploitation and pollution in its adjacent waters; and from the precedent established by the Latin American states in their unilateral and joint actions on the matter.[11]

In terms of the ongoing negotiations in the SBC and the General Assembly, the two declarations were important for three reasons. First, they announced the formation of an important group of 14 states united around a strong coastal state position (unilateralism), and including all the South and Central American coastal states except Venezuela and Costa Rica. Within the Latin American group as a whole, only the landlocked states (Bolivia and Paraguay), the big oil producers (Venezuela and Trinidad and Tobago), and most of the Caribbean countries (Barbados, Jamaica, Cuba, and Haiti) failed to support the unilateralist position. This was nonetheless a major advance on previous Latin American attempts to forge a unified position on the law of the sea, and the signers of the declaration comprised the first group created with an obvious orientation toward a new comprehensive law of the sea conference.

Second, the two declarations constituted a big step away from seabed discussions and toward consideration of coastal state jurisdiction. The 14 Latin American states, while not committing themselves to a specific form of jurisdiction, made explicit their determination to control exploitation of living and non-living resources, scientific research, and pollution as far off their coasts as they deemed necessary, but without prejudice to freedom of navigation. This effectively resumed the coastal versus maritime state dispute, and marked the beginning of a diplomatic offensive by the Latin American states to find allies on a strong coastal state position. In this context it is interesting to note that observers from Canada, Iceland, India, South Korea, Senegal, Egypt, and Yugoslavia attended the Lima meeting, and that a representative from Latin America was sent to the January 1971 session of the Asian-African Legal Consultative Committee (AALCC) to present the coastal state case.[12]

The third important aspect of the Lima and Montevideo Declarations was their implication for the boundary debate in the SBC. The Latin American 200-mile states had always been in the forefront of those opposing the early definition of a boundary for the international seabed area, and these two declarations institutionalized that opposition. By adopting an open-ended unilateralist position, the 14 Latin American states added their considerable weight to help perpetuate the dilemma of defining a boundary without knowing the nature of the machinery, or defining machinery without knowing the limits of the area to which it would apply. This perpetuation served their purposes admirably by providing time in which to build support for the higher priority coastal state interests.

A second important external development on the boundary question was the elaboration of a firm U.S. policy on the seabed in a statement by President Nixon on May 23. The Nixon statement was a landmark in several respects. On the boundary question the President proposed a three-level system in which the coastal state would have complete control over seabed resources within the 200-meter depth line, with all resources beyond that line becoming part of the common heritage. However, the area between the 200-meter line and the edge of the continental margin would be classified as an international trusteeship zone, within which the coastal state would retain many of its powers in return for sharing revenue from the zone with the international seabed authority. The authority would have exclusive jurisdiction only over the seabed area beyond the margin. Furthermore, the President proposed that until an international regime could be established, states should have the right to issue permits for exploitation beyond 200 meters, providing such permits were subject to the future regime, and provided also that the issuing state share "a substantial portion of the revenues" with "an appropriate international development agency."[13]

The U.S. proposal was, not surprisingly, in stark contrast to the Latin American declarations. It emphasized resources where the Latin Americans concentrated on territory; it stressed fixed boundaries where the Latin American states preferred ambiguity; it favored establishing boundaries before defining the regime and machinery where the Latin American states preferred the reverse; it confined itself rigorously to seabed issues where the Latin Americans were anxious to claim rights in the water column; it represented an attempt to compromise between wide- and narrow-shelf positions where the Latin Americans stuck with an extreme wide-shelf position; and it represented an attempt to circumvent the moratorium resolution by an interim policy where the Latin Americans had almost unanimously supported the moratorium resolution. Like the Latin American declarations, however, the U.S. proposal helped turn attention away from the international area issue and toward the problem of coastal state jurisdiction.

The proposal represented not only an attempt to offer an international compromise, but also an attempt to draw together the conflicting interest groups within the United States. The oil industry, speaking through the National Petroleum Council and the Department of the Interior, had made its position clear in 1969.[14] It wanted the government to "assert . . . its exclusive jurisdiction over the mineral resources of the entire submerged portion of the Continent off its shores." The Defense Department took the opposite view, arguing in favor of narrow-shelf limits on the grounds that wide-shelf claims would encourage wide claims over the water column, which in turn would prejudice freedom of navigation in ways detrimental to U.S. military interests. The State Department acted as mediator on the issue, and was largely responsible for working out the position adopted in the Nixon statement. It drew substantially on ideas developed by the Commission on Marine Science, Engineering, and Resources in its 1969 report "Our Nation and the Sea",[15] but it failed to win over the oil industry, which denounced the United States' proposal as a giveaway.[16] By way of imposing some coordination on the numerous bodies within the government interested in marine affairs, an Interagency Law of the Sea Task Force was established in February 1970 to deal with problems arising out of preparations for the third Law of the Sea Conference.[17]

The Nixon statement was an important contribution to the debate on a boundary for the area not only because it laid down a concrete American position on the subject, but also because it introduced a specific proposal into a debate that was much in need of something to give it focus. Considering the advanced state of oil exploitation off the coasts of the United States it was also a remarkably generous proposal, since much of the revenue to be shared from exploitation beyond 200 meters would come from the United States. The statement was formally introduced into the SBC on May 25, and reappeared in much more elaborate form in the draft convention on the international seabed area (138/25) submitted to the SBC by the U.S. delegation on August 3.

The two Latin American declarations and the U.S. proposal were the most important, but not the only, contributions to the boundary debate during the 1970 SBC sessions. Norway and Iceland both suggested variations on a 200-mile/500-meter distance/depth formula, while Canada and Italy continued to urge a boundary based on the continental margin. All of these proposals favored a generous allocation of the seabed to the coastal states, and taken together they suggested the extent to which discussion of the deep seabed had aroused coastal states to an awareness of the potential riches off their shores.

But despite this trend toward wide coastal state jurisdiction, there was no trend toward agreement on the boundary question. Britain and Australia, with their burgeoning offshore oil interests, did not favor the 200-meter limit in the U.S. proposal, preferring instead a wider, deeper limit. Kenya led the attack against the U.S. idea of a trusteeship zone, and Malta inveighed against the unilateralist position of the Latin American states. Thus, although the boundary debate was finally beginning to focus around specific options, it did not after three years appear to be any closer to a resolution.

The dispute over international machinery also continued to occupy the limelight in the committee's deliberations. This was especially true during the committee's last round of meetings in August, because at the beginning of that round the United States, Britain, and France each submitted proposals on the international seabed regime (respectively 138/25, 138/26 and 138/27). These proposals were the first in a series, and more properly form the subject of a subsequent chapter. Their major effect on the 1970 discussions was to act as a focus for debate in much the same way as the Latin American declarations and the Nixon statement had helped to crystallize the boundary debate.[18]

Disagreement centered on three questions, the first of which was the ongoing dispute over operating machinery in which the split between developed and developing countries continued to be the rule. The Soviet Union stuck to its position that priority should be given to international cooperation in ocean research, and that expectations of rapid exploitation were premature. Britain, Canada, Italy, and Australia again pointed out the danger of large machinery absorbing by itself most or all of the revenue generated from seabed mining, and the U.S., British, and French proposals all outlined a weak licensing type of international machinery. Japan also opposed operating machinery as impractical, but conceded that such machinery might be evolved in the future. On the other side, Kuwait outlined a strong machinery combining direct operations by the authority with a strong licensing format. The combination proposal represented some movement in the debate, as did the U.S. shift from favoring a registry regime to advocating a weak licensing type, but the full extent of alignments on these new proposals did not become clear before the end of the SBC's 1970 sessions.

The second and third disagreements both arose out of the U.S. position as defined in the Nixon statement and in 138/25. The suggestion for an interim regime predictably aroused the opposition of the Latin American states because of its implication for the moratorium resolution. The strict interpretation given to the common heritage, namely that it constituted the resources of the area rather than the area itself, elicited protest from a number of developing countries, including Chile and Trinidad and Tobago, on the grounds that it emasculated the common heritage as they understood it. As an effort at compromise, the U.S. proposal was clearly headed toward failure by the end of the 1970 SBC sessions.

Despite the controversy surrounding the international machinery question, the Economic and Technical Subcommittee managed to make considerable progress during 1970 in its discussion of the problems of exploitation. Although the machinery dispute spilled over into the subcommittee's work in the form of working papers by the United States and by 12 developing countries, the subcommittee managed to avoid becoming overly bogged down in it by concentrating its attention on the economic and technical problems facing operators under a variety of regimes. No agreement was reached, but the discussion was notable in the context of the SBC for its concreteness and depth of detail. An important base was laid down in these discussions for giving realistic consideration to the multitude of regime proposals that were beginning to appear before the committee.

The year 1970 was the last in which the SBC met purely to consider the questions raised by Pardo in 1967. Like the sessions of the previous two years, the 1970 session was inconclusive. The major objective of a declaration of principles remained beyond reach, and although a great deal of useful clarification had resulted from the year's work, no apparent progress had been made toward agreement on the most important questions. Even more significant than the committee's lack of progress was the increasingly obvious extent to which states were turning their attention toward wider issues of the law of the sea. In no small part this development was due to the increasing complexity and paralysis of the international seabed regime debate.[19]

Shortly after the end of the SBC's sessions another regional event of some importance to seabed politics occurred. This was the Lusaka Statement on the Seabed adopted by 53 nonaligned countries on September 10. The Lusaka Conference as a whole was concerned primarily with third world topics such as the entry of China into the UN, colonialism and racism in southern Africa, and the preservation of the nonaligned coalition. The fact that it concerned itself with seabed matters at all amounted to a significant comment on the importance attached to the subject by the developing countries.

The statement, unlike the Latin American declarations, was not of particular interest as a factor in seabed politics: for the most part it was a bland declaration of principles few of which would have raised eyebrows in the Seabed Committee. In terms of the active controversies in the SBC, it came down in favor of the common heritage defined as the area and its resources, powers to control mineral price fluctuations for the authority, and a comprehensive agenda for a new law of the sea conference. The nonaligned countries said nothing of interest regarding the machinery debate, and appeared to favor defining the regime before drawing a boundary for the area. All of these were positions already well established by the developing countries in the SBC debates.

What was significant about the Lusaka Statement was the group of countries adopting it. Twenty-six were African, 13 Arab, nine Asian, only four (Guyana, Jamaica, Trinidad and Tobago, and Cuba) Latin American, and one (Yugoslavia) European. In addition, 13 of the states were landlocked and six shelflocked; four had major interests as metal exporters (Congo, Uganda, Zaire, Zambia), while none had importing interests above a very low level. None were adherents to the Lima Declaration.

This group was clearly a new phenomenon in seabed politics, representing quite different interests from the other regional or political groups. It was African-dominated inasmuch as 32 of its members (60 percent) also belonged to the OAU, and the influence of copper-exporting states within the group was manifested by the reference to price controls in the statement. The inclusion of a large group of landlocked states, and the consequent orientation of the statement toward international area questions, was in marked contrast to the Latin American declarations, which had no landlocked adherents and were oriented toward coastal state concerns. The Lusaka Statement, in this sense, could be taken as an announcement that new interest groups were emerging as actors in the seabed and wider law of the sea discussions.

When the seabed debate moved again to the General Assembly in the autumn of 1970 it carried with it the agreement reached in the SBC to try to iron out the remaining difficulties barring completion of the draft declaration of principles. Pressure to complete the declaration in 1970 was considerable. Not only had the General Assembly requested such completion, but also the increasingly detailed level of discussion on both seabed and other law of the sea matters was leaving the general discussion of principles behind. Most major areas of dispute had become sufficiently defined that they could be incorporated in a declaration of principles, and this was done during the autumn 1970 sessions of the General Assembly's First Committee.

The declaration of principles that resulted was a statement of the maximum compromise possible after three years of negotiations.[20] In order to achieve agreement within the time limit set by the Assembly's agenda, it was necessary for some states to make concessions and for some issues to be left out of the declaration. Inclusion of the common heritage principle and a commitment to international machinery were essential if the declaration was to have the support of the great majority of developing countries, and this inclusion required concessions by the developed states, mostly from the Soviet group, that still maintained reservations on these two items. On the other side, the developing countries had to settle for a declaration that made no direct mention of the moratorium, and did not include control of metal price fluctuations as a principle, but only mentioned it in the preamble. Article 13a, which stated that the declaration did not affect the legal status of the waters superjacent to the area, can be seen as part of the process of compromise, in that it eliminated the possibility of the seabed machinery extending its jurisdiction over high seas fishery resources. Fear of this possibility was an important factor in the opposition to international machinery of some developed states, particularly those with major distant water fishing interests like the Soviet Union and Japan.

Most of the conspicuous deletions from the declaration of principles involved items that were under dispute within the framework of the general principles. These included operative definitions for "the common heritage" and "peaceful purposes" and for the structure and functions of the international regime and machinery. The transfer of technology was not mentioned explicitly, although there was mention in Principle 10c of strengthening the research capabilities of the developing countries. The major absence from the declaration was the whole question of defining limits for the area, and whether or not to define such limits before or after defining the international regime and machinery. This question had occasioned a heated controversy in the First Committee, with a dozen developing countries insisting that the regime must be defined before the boundary, and half a dozen developed states, including the Soviet Union, insisting that priority be given to establishing the boundary.[21] As a result of this impasse, the absence of limits was simply noted in the preamble to the principles with the comment that they were "yet to be determined." Thus the most crucial question before the delegates, and the one most paralyzing to the negotiations, was left unresolved.

The final draft of the declaration of principles was not a happy compromise, and showed signs of having been forced to a conclusion under pressure of time. The most obvious sign, other than the omissions mentioned above, was the burden of concession placed on the Soviet Union and its allies. The Soviets not only had to swallow the common heritage idea and the commitment to international machinery, but they had to accept the irresolution of the boundary issue, upon which they had campaigned hard for giving priority to determination of the area over definition of the regime. As a result of this burden none of the Soviet group states was among the 46 sponsors of Resolution 2749, which contained the declaration of principles, and nearly all of them were among the abstainers in the 108 to 0 to 14 vote in the General Assembly.

The Soviets did not stand alone in their dissatisfaction with the declaration of principles, for none of the other major developed countries joined the list of sponsors either. Among the Western states only those with large coastal state interests (Canada, Australia, Iceland, and Norway) were prominent among the sponsors. The absence of all the major powers and all but two landlocked countries (Austria and Lesotho) from the list of sponsors of what was supposed to be a key compromise resolution bespoke the tenuous nature of the agreements on which the declaration of principles rested.

Resolution 2749 was in many respects a culmination of the process started by Pardo in 1967. It was the final product of the three years' work of the Seabed Committee set up as a result of his initiative, and embodied all the most important points of his original proposal. As Pardo had suggested, the international seabed area had been declared the common heritage, and the principles of non-appropriation by states, use of seabed revenues to promote development, reservation of the area for peaceful purposes, and establishment of an international regime and machinery, had all been accepted.

While this accomplishment superficially appeared to be a success for Pardo, the underlying reality looked much more like failure. Although the principles were accepted, the ability to put them into operation was blocked by profound disagreements on interpretation. In addition, since even the first level agreement on principles had taken so long to obtain, the race for deep sea resources which Pardo had hoped to preempt was already well under way. This race, and the debate on the deep seabed, had acted to undermine further Pardo's initial hopes by reopening numerous related law of the sea questions. By 1970 these had acquired a momentum of their own that had almost certainly destroyed any possibility that the seabed issue could be settled on its merits as a separate item. For better or worse the seabed issue was, by 1970, inextricably tied to the whole review of the law of the sea.

In addition to the declaration of principles, the General Assembly also passed three other resolutions, 2750A-C, at its December 17 meeting. Resolution 2750A requested a study and proposals for solution of "the problems arising from the production of certain minerals from the area beyond the limits of national jurisdiction" in the light of "the impact they will have on the economic well-being of the developing countries, in particular on prices of mineral exports on the world market." The resolution reiterated the non-principle that the international regime should protect exporters against price fluctuations, and in general was symptomatic of the rising concern among the developing countries that seabed exploitation might prove more of a threat than a boon to their economic interests. It was sponsored by 17 developing countries, among which were four states with large copper exporting interests (Chile, Peru, Zambia, and Zaire); with the Lusaka Statement it was further evidence that the nodule-metal-exporting developing countries were mobilizing as an interest group. Japan and the Soviet Union both opposed the request for such a study in the discussions for it clearly had implications unfavorable to their interests—encouraging rapid exploitation of the seabed in the case of Japan (a large importer), and discouraging large and powerful international organizations in the case of the Soviet Union. The resolution nonetheless passed 104 to 0 to 16.

Resolution 2750B was another request for a study, this time on "the question of free access to the sea of land-locked countries . . . with a report on the special problems of land-locked countries relating to the exploration and exploitation of the resources of the sea-bed . . . beyond the limits of national jurisdiction." Twelve landlocked countries, all but one (Austria) developing, and two-thirds of them African states, sponsored the resolution, which passed in the General Assembly by a vote of 111 to 0 to 11. This resolution also carried implications of a newly emergent interest group.

Resolution 2750C was much the most important of the three, and marked a turning point in both the seabed debate and the history of the SBC. It followed on from Resolution 2574A of the previous year, which had requested the Secretary-General to make a survey of opinion on holding a comprehensive conference on the law of the sea. The results of the survey indicated widespread support for such a conference, as had the Lima Declaration and the Lusaka

Statement, and so in Resolution 2750C the General Assembly decided to

> convene in 1973 . . . a conference on the law of the sea which would deal with the establishment of an equitable international regime—including an international machinery—for the area and the resources of the sea-bed and the ocean floor, and the subsoil thereof, beyond the limits of national jurisdiction, a precise definition of the area, and a broad range of related issues including those concerning the regimes of the high seas, the continental shelf, the territorial sea (including the question of its breadth and the question of international straits) and contiguous zone, fishing and conservation of the living resources of the high seas (including the question of the preferential rights of coastal States), the preservation of the marine environment (including, inter alia, the prevention of marine pollution) and scientific research.

In order to prepare for such an ambitious conference the General Assembly reshaped the SBC into a preparatory body. This involved an enormous change in the SBC's mandate, so that its main task became that of defining a list of all the subjects and issues relating to the proposed conference and then preparing draft articles on them. Obviously the deep seabed issue was much more advanced than the other areas because of its three-year head start. Most of the task of defining its subjects and issues had been completed with the declaration of principles.

The transition to a preparatory body also involved a great increase in size for the SBC. Its membership of 42 was more than doubled by the addition of 44 more states. Thirty-four of these new members were developing countries (nine Latin American, 11 African, six Arab, eight Asian), and only nine were developed states (six Western and three Soviet group).[22] Landlocked states continued to be proportionally underrepresented even though four more were added. This huge influx of states was made necessary by the high level of interest in, and concern about, law of the sea questions among all groups of states. No state with any material interest in law of the sea matters wanted to be left out of negotiations that might radically affect its interest. Its effect was to bring the SBC more into line with the membership pattern of the General Assembly by increasing the ratio of developing to developed members. The SBC's consensus rule largely mitigated the political impact of the change in balance, but the increase in size nevertheless had two important effects. First, it made the work of the committee twice as cumbersome because of the larger number of speakers and the greater difficulty of achieving consensus. Second, it revived the educational aspects of the committee's work, since many of the new members had to acquire the necessary informational background.

Resolution 2750C passed by a vote of 109 to 7 to 6,[23] with all of the opposing and half the abstaining votes being cast by the Soviet Union and its allies. The Soviet Union remained opposed to a comprehensive conference, preferring instead an agenda restricted to territorial sea and continental shelf limits, straits navigation, and coastal state fishing rights. The United States, which had not originally supported the comprehensive conference approach, changed its position, even going so far as to join the list of sponsors of 2750C. This placed the United States firmly on the side of those wanting to negotiate the whole law of the sea as a package, in which conflicting interests in one area might be resolved by trading off concessions in other areas.

These changes in mandate and membership altered the SBC almost beyond recognition from its 1968-70 appearance, and in this sense Resolution 2750C closed the first round of international activity leading up to the Third UN Conference on the Law of the Sea, and opened the second. The first round stretched from the Pardo proposal in August 1967 to the passing of the declaration of principles in December 1970. It was motivated and dominated by seabed politics, and by the end of it the two major questions and some of the major alignments of seabed politics were quite firmly defined. The question of whether the international machinery should be strong or weak in regard to the process of exploiting the deep seabed was the central problem of the international seabed regime issue, and alignments on it cut clearly along developed/developing lines. The question of whether coastal state jurisdiction over the seabed should be narrow or wide in extent was equally important, but options and alignments on it were less fully developed.

By the end of the round, however, the whole context of seabed politics had changed. Instead of being considered on their own merits, seabed questions had become entangled in the wider coastal versus maritime interest conflicts of ocean politics. The much broader issues thus opened up tended to draw attention away from the complexities and uncertainties of the international seabed question and toward the more immediate, tangible, and emotional questions of coastal state jurisdiction. Coalitions were already forming on this broader basis, and related regional activity was emerging as an important counterpoint to the activity in the United Nations.

By the end of 1970, then, the relatively limited objectives of Pardo's proposal had been absorbed into a much more grandiose framework. As President Nixon phrased it, the issue at hand was now

> whether the oceans will be used rationally and equitably and for the benefit of mankind or whether they will become an arena of unrestrained exploitation and conflicting jurisdictional claims in which even the most advantaged states will be losers. . . . The stark fact is that the law of the sea is inadequate to meet the needs of modern technology and the concerns of the international community. If it is not modernized multilaterally, unilateral action and international conflict are inevitable.[24]

NOTES

1. GAOR XXIV, Suppl. 22, A/7622, 1969 (Report of the SBC), pp. 29-31.

2. C. B. Bourne and L. G. Jahnke, Cases and Materials on Public International Law (Vancouver: University of British Columbia, 1972), pp. 7-58 and 7-61.

3. The chronology on this matter was as follows. June 1968, United States and Soviet Union present draft resolutions to the AHSBC; July 1968, discussions begin in ENDC and move to CCD; March and May 1969, Soviet Union and United States draft treaties; October 1969, U.S./Soviet joint draft submitted to CCD; November 1969, SBC considers the joint draft; April 1970, United States and Soviet Union present revised draft to CCD, further negotiations and consultations over summer 1970; December 1970, UNGA accepts treaty in Resolution 2660 (104 to 2 to 2); February 1971, open for signature. The treaty provided that "The States Parties . . . undertake not to emplant or emplace on the seabed and ocean floor and in the subsoil thereof beyond the outer limit of a seabed zone (of 12 nautical miles) . . . any nuclear weapons or any other types of weapons of mass destruction as well as structures, launching installations, or any other facilities specifically designed for storing, testing, or using such weapons." See E. D. Brown, Arms Control in Hydrospace: Legal Aspects (Washington: Woodrow Wilson International Center for Scholars, 1971); Evan Luard, The Control of the Sea-bed (London: Heinemann, 1974), pp. 97-112; Louis Henkin, "The Sea-Bed Arms Treaty—One Small Step More," Columbia Journal of Transnational Law 10, no. 1 (1971), pp. 61-65; Lay et al., New Directions in the Law of the Sea (Dobbs Ferry: Oceana, 1973), pp. 288-91; William Butler, The Soviet Union and the Law of the Sea (Baltimore: Johns Hopkins Press, 1971), pp. 158-60; UN Doc. A/AC.135/28, July 1968; W. W. Krieger, "The UN Treaty Banning Nuclear Weapons and Other Weapons of Mass Destruction on the Ocean Floor," Journal

of Maritime Law and Commerce 3, (October 1971), pp. 107-28; and S. M. Selzer, "The Seabed Arms Limitation Treaty," Journal of Law and Economics 6 (June 1971), pp. 157-74.

4. On the SBC's work generally in 1969 see its report, op. cit.; and A/AC.138/SR.1-16; A/AC.138/SC.I/SR.1-29, and A/AC.138/SC.II/SR.1-25.

5. UN Monthly Chronicle 7, no. 1 (January 1970), p. 76.

6. This was not a recorded vote, but the alignments were similar to those on Resolution 2467C the previous December. At this time, both the United States and the Soviet Union favored conferences with tightly restricted agendas, and as early as 1967 the Soviet Union had begun exploring the possibility of a conference on the territorial sea and straits. Even then, however, the pressure to bring in more issues, particularly fisheries, was irresistible. Private discussions on the possibility of a conference on some law of the sea issue continued separately from, but parallel to, the work of the SBC through 1968-69, and the two streams were finally drawn together in the General Assembly battle over Resolution 2574A. (Interviews with delegates.)

7. This was not a recorded vote, but most of the abstentions were Soviet group states.

8. U.S. Senate, Mineral Resources of the Deep Seabed, 93rd Congress, first session on S.1134, p. 517.

9. GAOR XXV, Suppl. 21, A/8021, 1970 (Report of the SBC), Annex IV, pp. 127, 129.

10. A/AC.138/SR.44, pp. 175-77. The proliferation of small negotiating and drafting groups was an issue that occurred frequently during the law of the sea negotiations. The use of such groups to facilitate negotiation and drafting was a traditional method in international negotiations, and was favored by most developed states. The developing countries tended to oppose it for several reasons. First, it created problems for small delegations that did not have enough members to go around. Second, it weakened the cohesion of the developing group by dividing them into "ins" and "outs." Third, it undermined the majority strength of the Group of 77. And fourth, it meant that no records were available for much of the negotiating process.

11. The texts of both declarations are reprinted in Oda, The International Law of the Ocean Development (Leiden: Sijthoff, 1972), pp. 347-55.

12. E. D. Brown, "The 1973 Conference on the Law of the Sea: The Consequences of Failure to Agree," in L. M. Alexander (ed.), The Law of the Sea: A New Geneva Conference (Kingston: University of Rhode Island, 1972), p. 9.

13. The text of the Nixon statement is reprinted in Lay et al., op. cit., pp. 751-52; and as A/AC.138/22. The United States had already granted exploitation leases extending down to 5,000 feet for oil and gas drilling. See Edward Brown, The Legal Regime of Hydrospace (London: Stevens, 1971), p. 18.

14. National Petroleum Council, Petroleum Resources Under the Ocean Floor, Washington, 1969; also summarized in Oda, op. cit., pp. 333-35.

15. Summarized in Oda, ibid., pp. 327-33. The commission, also known as the Stratton Commission and COMSER, was an outgrowth of the 1966 Marine Resources and Engineering Development Act. It suggested a registry type of seabed authority and an intermediate zone between national and international jurisdiction for the area between the 200-meter/50-mile line and the 2,500-meter/100-mile line.

16. In a supplement to its 1969 report, summarized in Oda, ibid., pp. 335-37.

17. On the subject of internal U.S. politics on law of the sea matters, see Ann Hollick, "U.S. Ocean Politics," San Diego Law Review 10, no. 3 (May 1973): 467-501; "Canadian-American Relations: Law of the Sea," International Organization 28, no. 4 (Autumn 1974): 755-80; and Edward Wenk, Jr., The Politics of the Ocean (Seattle: University of Washington Press, 1972).

18. The text of the Nixon statement contained substantial hints as to the U.S. view on the regime and machinery, but the position was not fully elaborated until the release of A/AC.138/25.

19. On the work of the SBC generally in 1970, see GAOR XXV, Suppl. 21, A/8021, 1970 (Report of the SBC); and A/AC.138/SR.17-44, A/AC.138/SC.I/SR.30-38, A/AC.138/SC.II/SR.26-40. For Pardo's view of the work of the UN and the SBC from 1967 to 1970, see Arvid Pardo, "Development of Ocean Space: An International Dilemma," Louisiana Law Review 31, no. 1 (December 1970), pp. 45-72.

20. The text of the declaration is reprinted in Oda, op. cit., pp. 44-46.

21. UN Monthly Chronicle 8, no. 1 (January 1971), p. 40.

22. Four places were allocated to the East European group, but one remained unoccupied because of the group's unwillingness to seat Albania.

23. The recorded vote was 108 to 7 to 6, but Pakistan stated that the recording machinery had missed its affirmative vote.

24. A/AC.138/22.

CHAPTER 6

THE STATE OF SEABED POLITICS AT THE BEGINNING OF NEW INTERNATIONAL NEGOTIATIONS ON THE LAW OF THE SEA

NATIONAL CLAIMS UP TO 1970

By the end of 1970, the pattern of national claims over the oceans had changed markedly from what it had been in 1967, and beyond recognition from what it had been in 1958. Table 6.1 shows that in comparison to the prevailing trends in 1960 (see Table 2.1), the pattern of territorial sea claims in 1970 was much more heavily oriented toward claims of 12 miles or more. No new three- , four- , or five-mile claims were made between 1960 and 1970, and while five more claims in the six- to 11-mile range were made, the total number of claims in that category dropped because of states abandoning six-mile limits for broader claims. In terms of seabed politics, the large shift toward a 12-mile territorial sea was interesting primarily as an indication of the new weight provided by the developing countries on law of the sea issues. The alignments on the territorial sea issue were still much the same as they had been in 1958, with the Western states dominating the three-mile camp, and the Soviet group and the developing countries supporting 12 miles.

The most important aspect of territorial sea claims for seabed politics, however, was the small but growing number of states with claims beyond 12 miles. Prior to 1960 only one very small state, El Salvador, had made such a claim, and the category was generally dismissed as an aberration. By 1970 11 states, some of which commanded very substantial stretches of coastline, had claimed the 12-mile limit. Many of these claims included areas of seabed beyond the edge of the geographical margin, and taken together they constituted an important circumvention of the flexible

TABLE 6.1

Territorial Sea Claims of 77 States to the End of 1970

Group	4-5 Miles	6-11 Miles[b]	12 Miles		13-199 Miles	200 Miles
Western	Finland Sweden Norway Iceland	Greece Italy Israel Spain Portugal South Africa[a]	Canada[a]			
Soviet			Bulgaria Soviet Union	Romania		
Latin American		Dominican Republic[a] Cuba	Jamaica[a] Trinidad and Tobago[a] Colombia[a] Venezuela	Honduras[a] Mexico[a] Guatemala		Argentina[a] Ecuador[a] Panama[a] Uruguay[a] Brazil[a] El Salvador
Arab		Tunisia[a]	Algeria[a] Yemen (S)[a] Yemen (A)[a] Iraq Egypt	Syria[a] Kuwait[a] Libya Saudi Arabia Sudan		
African		Ivory Coast[a]	Mauritania[a] Malagasy Republic[a] Togo[a] Somalia[a] Nigeria[a] Gambia[a] Kenya[a] Mauritius[a]	Ghana[a] Sierra Leone[a] Tanzania[a] Liberia[a] Dahomey[a] Senegal[a] Equatorial Guinea[a] Ethiopia	Guinea (130)[a] Cameroon (18)[a] Congo (15)[a]	
Asian		Yugoslavia[a] Sri Lanka	North Vietnam[a] Cyprus[a] Thailand[a] Burma[a] Malaysia[a] Indonesia China	Turkey[a] Pakistan[a] India[a] Cambodia[a] Albania[a] Iran	Philippines[a] (archipelago) Maldives[a] (rectangle)	
Total claims	4	12	50		5	6
Claims since 1960	0	5	36		5	5
Total claims up to 1960	5	14	16		0	1

[a]Since 1960.

[b]All these claims are for six miles except that of Yugoslavia, which is for ten.

exploitation limit laid down in 1958. Using the territorial sea loophole, states were claiming areas of seabed that were many years away from exploitation in almost any sense. The 11 states formed a group too substantial to be ignored, and their presence threatened to turn the already difficult three- versus 12-mile dispute into an almost impossibly complex three-sided one. It also threatened to undo the legal separation of the seabed from the water column so painstakingly achieved in 1958.

Table 6.2 amplifies the importance of the 12-mile-plus territorial sea claims for seabed politics. Compared with 1960 (see Table 2.2), the pattern of continental shelf claims had undergone an even more dramatic change than the pattern of territorial sea claims. Very few claims of the unspecified geological or fixed depth type were made between 1960 and 1970, and the membership of both categories declined. By contrast, a large number of states representing all regional groups accepted the flexible limits definition established by the 1958 Continental Shelf Convention. One-third of the states accepting the convention's definition did so without ratifying the convention itself, but this in no way detracted from the pronounced shift away from vague geographical definitions toward the exploitation definition.[1]

As with the 12-mile territorial sea, however, the major shift was not the one most important for seabed politics. The number of states making their continental shelf claims on the basis of fixed distance criteria nearly tripled in the 1960s, and all but one of these states made its claim within the context of a territorial sea or exclusive economic zone. Dahomey was the only state to lodge a direct continental shelf claim that went against the limits definition of the convention (discounting the three states that adopted the old unspecified geological approach).

The circumvention of the flexible limits compromise on continental shelf claims by claims to broad territorial seas and economic zones based on fixed distance lines would have been a significant development even if the Seabed Committee had not been created. Within the context of the SBC debate, however, it was a crucial event. That debate had generated an urgent requirement for fixed limits to coastal state jurisdiction over the seabed in order to define the boundary of the international area. Thus the SBC debates were undermining the 1958 convention settlement from the opposite direction, but at the same time, as the fixed distance national claims to broad territorial seas and economic zones. The result of this confluence was to leave the states that had backed the flexible definition in a weak position with regard to the SBC debate, while giving the fixed distance states a strong position. This was particularly true for the Latin American 200-mile states with their coordinated regional stand.

TABLE 6.2

Continental Shelf Claims of 100 States to the End of 1970

Group	Unspecified Geological	50 Meters	200 Meters or Variants	Exploitation with or without 200 Meters		200 Miles (fixed distance) or Variants
Western	Ireland[a] Greece[b] Spain[b]			Norway[a] Italy[a] Denmark[a, c] New Zealand[a, c] Netherlands[a, c] Australia[a, c] Portugal Canada[c] South Africa[c] Israel[b, c]	West Germany[a] Iceland[a] United Kingdom[a, c] Finland[a, c] Sweden[a, c] France[a, c] Switzerland[c] Malta[c] United States[c]	
Soviet				East Germany[a] Romania[c] Czechoslovakia[c]	Soviet Union[a, c] Poland[c]	
Latin American	Guyana[b] Bahamas[b] British Honduras[b] Falkland Islands[b]			Argentina[a, e] Uruguay[a, e] Jamaica[c] Trinidad and Tobago[c] Guatemala[c] Venezuela[b]	Brazil[a, e] Dominican Republic[a, c] Mexico[c] Colombia[c] Haiti[c] Honduras[b]	Nicaragua[a] Chile[b] Peru[b] Costa Rica[b] Ecuador[b, d] Argentina[a, d, e] Panama[a, d] Brazil[a, d, e] El Salvador[b, d] Uruguay[a, d, e]
Arab	Oman[a] Saudi Arabia[a] Libya[b] Iraq[b] Kuwait[b] Bahrein[b] Qatar[b] Seven United Arab[b]			Syria[a] Yemen (A)[a] Egypt[b]	Sudan[a] Yemen (S)[a] Morocco[b]	
African			Ivory Coast[a] Ghana[a]	Nigeria[a, c] Mauritius[a, c] Senegal[c] Malagasy Republic[e] Uganda[c]	Liberia[c] Malawi[c] Kenya[c] Sierra Leone[c] Swaziland[c]	Dahomey (100)[a] Guinea (130)[a, d]
Asian	Iran[b] South Korea[b] Sri Lanka[b] Burma[b]		Pakistan[b]	Malaysia[a, c] South Vietnam[a] Albania[c] Yugoslavia[c] Taiwan[e]	Philippines[a] Indonesia[a] Cambodia[c] Thailand[c] India[b]	Philippines[a, d, e] Maldives[a, d]
Total claims	25	0	3	62 (42[c]) -4 58	Double listed	14
Claims since 1960	3	0	2	56		9
Claims up to 1960	33	1	5	6		5

[a]National claim made since 1960

[b]National claim before 1960

[c]Ratified continental shelf convention

[d]Territorial sea claim

[e]State listed in two places

Source: Data derived from a wide variety of sources, the most comprehensive of which is U.S. Department of State, Office of the Geographer, Bureau of Intelligence and Research, Limits in the Seas No. 36: National Claims to Maritime Jurisdiction, April 1974.

The great increase in national claims, both territorial sea and continental shelf, between 1960 and 1970 was accompanied by increases in both the number of agreements and the number of disputes relating to delimitation problems. Up to 1967, delimitation agreements were concentrated largely in the North Sea, with only a few additional signs of activity in the Persian Gulf and the Baltic. By the end of 1967, in fact, the North Sea shelf was completely divided up among the six countries bordering it with the exception of the small area still under dispute at the ICJ between West Germany on the one hand and Denmark and the Netherlands on the other. The delimitations were done mostly on median line principles, and they applied only to the seabed, not to the superjacent waters. Britain and Norway both acquired very large tracts of seabed under these agreements, the border between them stretching up to 170 miles from the British coast.

After 1967, the focus of delimitation activity shifted away from the North Sea, but remained concentrated in relatively small, enclosed bodies of water surrounded by several states—particularly the Baltic Sea and the Persian Gulf. In the Baltic there were shelf delimitation agreements between the Soviet Union and Finland, Norway and Sweden, East Germany and Poland, and Poland and the Soviet Union, and territorial sea agreements between Norway and Sweden, and Finland and the USSR. In the Persian Gulf there were shelf delimitation agreements between Saudi Arabia and Kuwait, and Iran and Kuwait, and territorial sea agreements between Saudi Arabia and Qatar, Abu Dhabi and Dubai, Iran and Saudi Arabia, Abu Dhabi and Qatar, and Iran and Qatar. These delimitation agreements did not result in complete apportionment along North Sea lines. During this period also there were delimitation agreements on the shelf between Italy and Yugoslavia, and Indonesia and Malaysia, and on the territorial sea between Brazil and Uruguay, and Indonesia and Malaysia. As with the North Sea, many of these agreements were motivated by the worldwide expansion of offshore oil exploration and exploitation.

A counterpoint theme to delimitation agreements was the growth of delimitation conflicts that accompanied the increase in seabed claims. These too were largely related to oil interests, and in contrast to the situation in 1958, they added a new ingredient to marine boundary disputes that had traditionally been centered on fishing interests. Seabed boundary disputes did not develop into an important influence on seabed politics at the international level during the 1960s, but it was becoming clear going into the 1970s that a considerable number of potentially serious disputes were on the horizon. The progress of the Seabed Committee and the spread of the offshore oil industry to deeper and more far-flung sites were both factors in stimulating conflicting claims. The SBC debates made it clear that the whole seabed was about to be allocated in the relatively near future, and this encouraged states to cast their attention further out than they would otherwise have done. The offshore oil industry stimulated hopes of wealth for coastal states because of the frequency of its finds, and its increasing ability to drill in deeper waters encouraged states to claim the largest possible shelf area in order to maximize their chances of an offshore bonanza.

East and Southeast Asia were the areas most productive of disputes despite early efforts to prevent conflicting claims. In 1966 the UN Commission for Asia and the Far East (ECAFE) established a Committee for Coordination of Joint Prospecting for Mineral Resources in Asian Offshore Areas (CCOP). The CCOP included Japan, South Korea, the Philippines, and Taiwan among its original members, and later Cambodia, Indonesia, Malaysia, South Vietnam, and Thailand. Its purpose was to forestall conflict through joint activity, but an extensive prospecting survey sponsored by the CCOP in 1968 produced an optimistic assessment of oil potential, and this precipitated a rash of conflicting claims.[2]

Cambodia, Thailand, and South Vietnam all pushed out extensively overlapping claims into the Gulf of Siam,[3] and South Korea, Japan, Taiwan, and China made conflicting claims to large areas under the Yellow and East China Seas, thereby adding yet another element into the tangled and conflict-prone relationships surrounding that area. In addition, China and South Vietnam disputed ownership of islands in the South China Sea.[4] This latter conflict exemplified the difficulties created in seabed delimitation by the existence of disputes over the ownership of islands—a problem that was to appear in many other places and that was to give unprecedented importance to previously insignificant rocks and reefs. Potential disputes along similar lines also existed among Malaysia, Indonesia, the Philippines, China, and South Vietnam in the South China Sea.

In the area as a whole, the problem was complicated by the high demand for oil from Japan, Taiwan, and South Korea, all of which were heavily dependent on foreign supplies, and by the awkward diplomatic position of China in relation to the other states. Late in 1970 the Chinese reacted strongly to the delimitation rush that had been going on off their coast by voicing firm support for the Latin American claims, denouncing the development efforts in the Yellow, East China, and South China Seas, and lodging their own claim to sovereignty over continental shelf resources. This Chinese response quickly dampened hopes for early exploitation in the area.[5]

Latin America also harbored several disputes, though none of them were prominent before 1970. Colombia and Venezuela were unable to settle their boundary off the entrance to Venezuela's oil-rich Gulf of Maracaibo. Chile, Peru, and Bolivia entertained a long-standing dispute relating to the war settlement of nearly a century before, in which Peru had lost coastal territory to Chile and Bolivia had lost its coastline entirely. The Bolivians in particular were unhappy about their landlocked status, and the exploitation of oil off the coast of Peru did not diminish their ardor for rectification of the injustice. Argentina and Britain had a long-standing dispute over the Falkland Islands, which, given the location of the islands in the middle of the extensive Argentine continental shelf, could not help becoming increasingly controversial as interest in offshore oil quickened.

In North America there were active but unsettled negotiations between Canada and Denmark over the boundary between Greenland and Baffin Island, between Canada and France over delimitation of the continental shelf around the islands of St. Pierre and Miquelon (just south of Newfoundland), and between Canada and the United States over two oil potential areas: the Georges Bank between Maine and Nova Scotia and their boundary in the Beaufort Sea.

In Europe, seabed delimitation disputes existed at the two geographical extremes of the continent. In the far north, Norway and the Soviet Union faced a difficult problem in delimiting their boundary on the deep but potentially oil-rich shelf under the Norwegian and Barents Seas. The problem was compounded by the ambiguous status of the Spitzbergen Islands and by the strategic significance of the area for the Soviet navy. In the far south the perennially fractious relationship between Greece and Turkey was aggravated further by the seabed delimitation issue. Greek islands sat very close offshore from the Turkish mainland, and were in fact located on the shelf that was a geographical extension of Turkey. Oil interest was alive in the Aegean as elsewhere, and no proposals existed that seemed likely to succeed in reconciling this conflict between political and geographical realities.

The problem of islands belonging to one country but located on the physical shelf extending from another was not limited to Greece and Turkey. The Falkland Islands and St. Pierre and Miquelon represented similar situations, and other examples existed as potential disputes in Italian islands off the coast of Tunisia, British islands off the coast of France, and Yemen (Aden) islands off the coast of Somalia.

The intensity of these disputes varied from mere recognition, as in the case of Britain and France, through polite negotiation, as in the case of France and Canada, to gunboat diplomacy, as in the case of South Vietnam and China. Their existence, and their threatening potential for escalation, provided an important motive for a new Law of the Sea Conference, and the prospect of such a conference in turn caused many of the disputing parties to act with restraint in anticipation of its outcome. The depth of feeling aroused by the prospect of offshore oil wealth was illustrated best by the disruptive impact it had on relations within as well as between states. The United States, Canada, and Australia all underwent substantial internal battles between federal and provincial or state governments over continental shelf resource jurisdiction, and in Britain offshore oil wealth fueled the fortunes of Scottish separatism. In terms of the upcoming conference, the disputes were symptomatic of the enormous difficulty inherent in designing a general settlement in the face of myriad local dilemmas, special cases, and unresolved conflicts over territorial claims. Perhaps more than anything else, the oil-motivated disputes over the continental shelf served as an example to the international community of President Nixon's 1970 warning that if the law of the sea were not modernized multilaterally, international conflict over unilaterally pronounced claims would be inevitable.

DEVELOPMENTS IN MARINE TECHNOLOGY AND OCEAN ACTIVITY, 1968-70

The massive expansion in the scale of ocean use and the rapid pace of innovation in marine technology that were discussed in Chapter 3 continued unabated through the later years of the 1960s. The world fish catch rose at a steady rate of between 2 and 3 million metric tons per year, so that by 1970 the total haul approached 70 million metric tons—a 75 percent increase over the 1960 total. The shipping industry likewise sustained its rate of expansion, and increased its move to very large bulk commodity carriers of questionable safety in order to serve the needs of the even more rapidly expanding world oil business. Oil pollution became the visible symbol of an increasingly articulate movement to protect the environment in general and the oceans in particular from the ravages of industrial civilization. And pressure to increase exploitation of the oceans on all levels continued to mount as more people in more states began to expect technology to provide them with an ever higher standard of living.

The extraordinary expansion of the offshore oil industry remained the most dramatic aspect of ocean activity through the later 1960s and on into the 1970s. By 1970 oil or gas had been discovered off the coasts of 28 states, and prospecting activity was under way on the shelves of 75 states.[6] Offshore oil was providing 17 percent of world oil production, accounted for 90 percent of the value of all seabed mineral production,[7] and was close to challenging fishing as the biggest source of ocean wealth. By 1970 two-thirds of the prospective acreage of shelf under less than 1,000 feet of water off the coast of noncommunist countries was already under license,[8] and it was commonly expected that offshore oil would provide more than 30 percent of world oil production by the end of the 1970s. American capital and technology retained their dominance in the industry, one author estimating that more than 15 billion dollars had been invested up to 1971, and that investment was proceeding at a rate of 3 billion dollars per year.[9]

The offshore oil industry was giving first priority in its exploration activity to the vast shelf areas around Australia, Indonesia, Malaysia, and Thailand, with West Africa and the North Sea competing for second place. The political problems caused by its expansion had not gone unnoticed within the industry, and it is interesting in this context to note an article in the Oil and Gas Journal[10] commenting on the problem of "unresolved boundaries" in relation to the industry's need to obtain exploration and exploitation permits. In addition to the ones discussed above, the author listed the following unresolved boundaries of concern: Sweden and the Soviet group countries in the Baltic; Iran, Iraq, and Kuwait in the Persian Gulf neutral zone; Australia and Indonesia in the Timor and Arafura Seas; Indonesia and the Philippines in the South China Sea; and Uruguay and Argentina in the mouth of the River Plate.

The industry continued to benefit from a high demand for its product and from improvements in its technology. Production moved steadily into waters more than 300 feet deep, and in 1970 Lockheed tested a deep sea production system designed to operate in depths up to 1,200 feet. Exploratory drilling stayed largely within the commercially interesting areas of less than 1,000 feet, and was concentrated on increasing the amount of deep drilling rather than setting new depth records. An indication of this was the growth in the number of rigs capable of deep water drilling—well over one-third of the mobile rigs operating in 1970 were capable of drilling in depths over 600 feet.[11] Experimental drilling rigs took cores bearing signs of oil from the seabed from over 11,000 feet down in the Gulf of Mexico during 1968. The phenomenal costs of offshore oil, with enormous profits as the prize, were illustrated by the $1.5 million price for a single North Sea well, and by the $850 million offered in high bids on 590,000 acres of shelf off Louisiana in December 1970.[12]

The other side of the offshore oil industry was the danger of environmental damage and pollution increasingly associated with it as it expanded. The supertanker aspect of the danger, which had been highlighted by the Torrey Canyon sinking in 1967, related to the oil industry as a whole, and not just to the offshore part. But the danger to the oceans from drilling activity itself did not have to wait long to find its own archetypal incident. In January 1969 an oil slick covering 800 square miles was unleashed from a seabed well off the California coast near Santa Barbara.[13] Despite precautions, oil flowed from the well for 11 days at a minimum of 1,000 gallons per hour,[14] at a distance of six miles from the coast. The Santa Barbara spill, though by no means the first of its kind, dramatized the threat to the environment generally, and to the fishing and tourist industries in particular, that was inherent in offshore oil drilling operations. It raised the question of coastal state control over such pollution should it occur outside the territorial sea, and generally tended to reestablish the link between the seabed and the water column in terms of coastal state responsibility. Offshore oil drilling thus encouraged coastal state interests by creating both the prospect of gain and the threat of costs.

The offshore hard minerals industry continued to place a poor second to oil in terms of global interest. The ongoing developments in nodule mining technology and activity (see Chapter 4) attracted a great deal of interest in the Seabed Committee discussions, but had become more of a political than an economic issue. Economic interest centered mainly in the United States, Japan, and West Germany, where work on the technology was in progress and there was some prospect of gain from opening up new sources of imported metals. For most other states the prospect of economic gain from nodule mining appeared very remote, and for many the threat of competition with their exports dominated attitudes. This was in complete contrast to offshore oil, which offered the prospect of large and direct benefits accruing to the coastal state in the relatively immediate future.

Hard mineral mining on the continental shelf maintained its steady, unspectacular growth but remained a backwater of interest. The Underseas Technology Handbook/Directory for 1973[15] listed 26 exploitation operations off nine states and 90 exploration operations off more than 25 states up to 1971. Activity in the industry was increasing on the basis of improvements in marine technology generally and increased demand for resources, but operations were still confined largely to waters under 200 feet deep, and neither the economic stakes nor the political difficulties involved approached those associated with the offshore oil industry. Continental shelf hard mineral mining thus remained a phenomenon of mostly local interest of the type generated by a brief scramble among U.S. companies during 1968 for rights to exploit the hot brine metalliferous muds in the Red Sea.[16]

ALIGNMENTS IN SEABED POLITICS

By the end of 1970, the debate on the seabed was sufficiently advanced that alignments of states had already begun to form on seabed issues. In some cases these alignments were well established around explicit policy questions; in others they were barely beginning to coalesce out of the major trends in the debate and had neither definite membership nor clearly defined positions. Since many of the important alignments in seabed politics were already visible in some form by the time the restructured SBC reconvened in 1971, it is worthwhile to examine them in more detail—particularly as to their strengths and weaknesses in terms of seabed politics.

Obviously not all these alignments were of the same type, and here a few general categories are useful. First is the distinction between alignments that were explicitly conscious of their own existence and alignments that emerged without conscious political organization or management. Although the dividing line between these types was sometimes blurred, the former could be distinguished by such activities as joint sponsoring, group declarations, and attendance at caucusing meetings, while the latter could be discerned by patterns of voting unaccompanied by such activity or by uncoordinated declarations, statements, and claims. Alignments acquired visibility through these activities, and part of the political process of the negotiations was to detect and trace these clues to underlying political movements. For our purposes here, alignments of the more self-aware type will be referred to as groups, while those of the more random type will be referred to as schools.

Second is the distinction between alignments that had their base outside the seabed and law of the sea negotiations and those that had their base within the negotiations. The former which can be called external alignments, consisted primarily of established regional groups, often with long-standing organizations such as the EEC and the OAU; political groups and schools such as the Group of 77; and the general alignment of developed states, which behaved mostly as a school but had some characteristics of a group in the Organization for Economic Cooperation and Development (OECD). The latter which can be called issue-specific alignments, consisted of groups and schools such as the landlocked states and the coastal states that owed their existence wholly to factors arising out of the negotiations. All of the external groups and schools correlated with the developed/developing dichotomy in the sense that all, or nearly all, of their members came from one camp or the other. This was much less true of issue-specific alignments, which frequently cut across developed/developing lines, creating unique coalitions.

In the early stages of negotiations on the seabed, between 1968 and 1970, the most visible alignments were external. At the time, seabed politics was most frequently characterized as polarized along developed versus developing lines, but as has been suggested in the preceding two chapters, this description was at best a very broad oversimplification. The remainder of this section will be devoted to an examination first of the developed/developing set of alignments, second of the regional alignments that both composed and divided the developed/developing set, and third of the various issue-specific alignments generated by the negotiations. It will conclude by relating these alignments to the condition of seabed politics at the beginning of 1971.

The developing countries nearly all belonged to the Group of 77, which by 1968 had become an established institution in global economic negotiations. The Group's objective was to reduce the economic disparity between developed and developing countries regardless of social system, and it was on this basis that it became a major actor in the seabed negotiations. Pardo's proposal to declare the seabed beyond the limits of national jurisdiction the common heritage of mankind was ideally tailored to the Group's purposes, and provided a strong position around which its members could rally. At the outset of the negotiations, the common heritage proposal appeared to offer an innovative and forward-looking opportunity to further the Group of 77's interests. If carried out, it would anticipate the opening of a new gap between developed and developing countries and provide for measures that would ensure utilization of new resources in a framework that would reduce rather than increase economic inequalities. An added advantage of the common heritage proposal was that its existence outside the geographical and legal framework of state sovereignty seemed to sidestep many of the normal difficulties confronting the Group. Not only would the proposal have to be both negotiated and implemented at the global level where the Group's voting power was most effective, but also the physically external nature of the deep seabed meant that differences of opinion within the Group should be less troublesome than on matters more directly related to state sovereignty.

In terms of seabed politics, then, the Group of 77 had several strengths. It was a solidly established entity with familiar procedures and objectives clearly relating to the issue at hand. It commanded a formidable block of votes in a forum where voting power counted. It faced its traditional opponents, and could therefore call on internal unity. And it was dealing with an issue both well suited and important to the furtherance of its interest. In addition, neither the Group as a whole nor many of its members individually had much to lose by taking an aggressive line in support of Pardo's proposals. Very few developing countries had much of a stake in the existing law of the sea, and all of them had something to gain if the common heritage could be put into operation.

Against these strengths, however, were a considerable number of weaknesses. The Group of 77 contained a large number of states representing a formidable variety of needs and interests. These included, for example, Kuwait in contrast to India and Tanzania in contrast to Haiti, as well as more general contrasts such as those between landlocked and coastal states, and between metal-exporting and metal-importing states. While the size of its membership was the essence of its strength, it meant that building a consensus on even the most general position became a major task, and that elaborating very detailed positions was almost impossible. Added to this difficulty was the fact that the Group's centrifugal tendencies were not random, but were themselves organized along regional lines. The Latin American, Arab, and African groups existed almost wholly within the context of the Group of 77, and represented alternate rallying points for members if consensus within the 77 became too distasteful. This factor was emphasized by law of the sea issues inasmuch as geographical features complemented regional alignments, as with the large coastal states of Latin America and the small coastal state/landlocked configuration in Africa.

Aside from the awkwardness of its internal politics, the Group of 77 suffered from a lack of knowledge, experience, and technical capability in relation to the seabed. None of its members had an independent capacity for deep sea drilling or dredging; only a few had oceanographic programs, and none of these even approached the level and scale of the major oceanographic powers; and few of them could afford to place acquisition of such skills very high on their list of development priorities. These deficiencies meant that the Group as a whole was poorly qualified to analyze the technical and economic aspects of seabed questions, and this characteristic reinforced the political tendency for the Group to confine its proposals to a very general level. These deficiencies, particularly when tied to the acute shortage of capital among all but a few oil-rich states, also meant that the Group lacked the power to establish an independent presence on the deep seabed. Since none of the Group's members had the capability to undertake nodule mining, and since none of them could exercise significant naval power over the high seas, they were placed in an essentially weaker position than the developed states.

The power of developing coastal states to make unilateral claims to the seabed was both a strength and a weakness for the Group of 77. It was a strength inasmuch as it gave the Group something to bargain with, but a weakness in that it created internally divergent commitments tending to undermine efforts to achieve unified positions. It is interesting to note in this regard that of the 13 most active developing countries in the law of the sea negotiations up to 1970, nearly half were states with major unilateral claims to coastal state jurisdiction (Sierra Leone, Sri Lanka, India, Brazil, Chile, and El Salvador).[17]

Given these limitations, the Group of 77 played a remarkably effective role in the seabed negotiations between 1967 and 1970. The developing countries with representatives in the Seabed Committee maintained a respectably solid unity on the great majority of issues throughout the debate on a declaration of principles, and assiduously exploited the UN system as a source of information to help compensate for their technical deficiencies. In the General Assembly, the developing countries grasped the initiative from the start and maintained sufficient (though far from perfect) unity to dominate the proceedings. The two resolutions calling for studies of international machinery (2467C and 2574C) and the resolution calling for consideration of a new, comprehensive law of the sea conference (2574A) provided the clearest examples of proposals sponsored by a large group composed exclusively of developing countries; they passed on the strength of the Group of 77's majority. Through the use of its voting power in this way, and because of its relative freedom from the constraints of vested interest in the law of the sea, the Group of 77 effectively directed the course of the seabed negotiations along paths favorable to its interests.

The moratorium resolution (2574D), by contrast, demonstrated the limits of the Group of 77's effectiveness both in terms of its own internal unity and in terms of its ability to wield influence outside the bounds of the General Assembly. This resolution was also symptomatic of a decline in the Group's effectiveness as a focus for alignment on law of the sea issues. As the discussion in the SBC and the General Assembly moved to more detailed levels and expanded to cover a wider range of topics, the Group's internal diversity began to come into play with greater force. On questions like the international seabed regime and machinery and scientific research, where interests were clearly divided along developed/developing lines, the Group of 77 managed to hold a strong and fairly unified position. But on questions like fixing the boundary between national and international seabed areas and determining the distribution of fishing rights, where geographical and economic characteristics assumed primary importance, the Group largely succumbed to internal differences. On this latter type of question alternative alignments such as those represented by the landlocked states, the metal exporting states, and the Latin American coastal states were already emergent by 1970.

No equivalent group to the Group of 77 existed on the other side of the developed/developing dichotomy. The Organization for Economic Cooperation and Development (OECD) counted most of the Western states among its members, but excluded the Soviet group states. In general the OECD states played opposite the Group of 77 in international economic negotiations, but the organization itself did not take a prominent role. Thus, in the sense of having an organized expression, the developed states alignment existed as a school rather than a group.

To a considerable extent, in fact, the developed states school existed only because the Group of 77, which included all the other states, defined its purpose as reducing the inequalities between developed and developing countries. In this sense, the developed states school was a negative alignment forced to respond to the purposeful initiatives of the Group of 77. The developed states school, however, did have more positive aspects. The great majority of its members had a substantial stake in the law of the sea as codified in the four Geneva Conventions. Their maritime interests and traditions predisposed them toward a system they had created to serve the needs of maritime states, and they had a long history of working together in the numerous international and transnational organizations concerned with marine affairs. Many of them also shared functional interests in shipping, distant water fishing, oceanographic research, and the exercise of naval power. In all these aspects they were largely separate from the developing countries, few of which had historical or functional stakes in the existing law of the sea.

Like the Group of 77, the developed states school also incorporated regional subgroups. These subgroups were much stronger in relation to the overall alignment than was the case among the developing countries; they included the Soviet Union and its allies, the western European states with the EEC members as their core, and the United States, which, due to the pluralistic nature of its internal politics, can in many ways best be treated as a group in its own right. The existence of a developed states school was demonstrated most clearly in the moratorium resolution, and less clearly in the large concentration of developed states among those states opposing or abstaining on Resolutions 2467C (international machinery study) and 2574A (consideration of a comprehensive new law of the sea conference).

The distinctive role of the Soviet group within the developed states school was illustrated by the frequency with which the Soviet Union, and some or all of its allies, could be found as the nearly exclusive occupiers of the abstained from or opposed law of the sea votes (Resolutions 2467A, 2467C, 2574C, 2750C). By comparison, the West had a relatively poor record of unity. Some of the small and medium-sized developed countries (like Norway, Iceland, Sweden, Canada, New Zealand, and Australia), often those with strong coastal state interests, tended to side with the developing countries on some issues. Only the major maritime powers could quite frequently be found voting together. Except for the breakaway group of developed coastal states, few developed states acted as sponsors of the important law of the sea resolutions, the major exception here being the United States on resolution 2750C (fixing a new comprehensive law of the sea conference).

The strength of the developed states school lay almost wholly in its economic and technological capability. Among themselves, the developed states controlled exclusively all the technology necessary for exploiting the seabed. In addition, they were responsible for the great majority of all ocean activity, and consequently controlled not only the hardware necessary for large-scale fishing, shipping, oceanographic research, and naval activity, but also most of the technological, scientific, and economic knowledge underlying such activity as well. In political terms, this strength gave the developed states a great deal of leverage in the debate on a new legal regime for the seabed and ocean. To a considerable extent, the international system as a whole was dependent on the maintenance of a high level of ocean activity—particularly trade carried by world shipping—and consequently any new legal regime would be obliged to attract the support, or at least the tolerance, of the developed states if it was to have any chance for success. An extension of this indispensability argument made it clear also that the developed states possessed a strong unilateral action option against any legal regime with which they disagreed.

Against this one strength, however, the developed states school had to contend with a substantial number of severe weaknesses. It lacked an overall coordinating and caucusing structure to compare with that developed by the Group of 77. This deficiency meant that it was less able than its opponent to counter the centrifugal pressures of divergent interests and alternative subgroup alignments, and consequently had to rely more on the uncertainties of common perceptions of interest to maintain unity. But even when it achieved unity, it still did not muster anywhere near enough votes to match those of the Group of 77. This situation forced the developed states to expend considerable amounts of political capital ensuring that the Seabed Committee, and later the Third UN Conference on the Law of the Sea, were conducted by consensus rather than majority vote.

But perhaps the greatest weakness of the developed states lay in the contrast between their outlook on the seabed and law of the sea renegotiation and the outlook of the developing countries. With relatively few exceptions, the developed states adopted a conservative, reactive attitude to the negotiations, as a consequence of which they largely forfeited their option to take the initiative. The reasons behind this appear to have been their extreme reluctance to abandon a regime favorable to their interests, and an underestimation of the strength of the developing states' reforming drive. Since the developed states possessed all the advantages in use of the oceans, they had a great deal to lose by changes in the legal regime. Instead of recognizing the historical imperatives that made some kind of radically new regime inevitable, most developed states roused themselves only to an unimaginative defense of their vested interests. In doing so they failed to appreciate the nature of the situation created in good part by their own excesses of ocean use. They allowed the developing countries, who had the stimulus of nothing to lose and everything to gain, to take the initiative, and consequently found themselves permanently on the defensive with regard to both the conditions and contents of the ensuing negotiations.

In summary, then, up to the end of 1970 the developed versus developing countries clash on the deep seabed issue can be characterized as a defensive mobilization of economic and technological power on the one hand, and an offensive mobilization of voting power on the other. Both sides were strong on this issue because the internal divergences among their members were least active, and the incentives for a united front were high.

In terms of the law of the sea negotiations the developed/developing country alignment as a primary framework was unsatisfactory and inefficient in several respects. To a certain extent it imported an institutionalized confrontation arising out of other issues into the negotiations. This tended to exacerbate an already difficult situation, and so decreased the likelihood of the negotiations resulting in the widely acceptable compromise necessary for a successful settlement.[18] Furthermore, the developed and developing country alignments were ill suited to many law of the sea issues in that they could not represent the interests concerned because of internal conflicts and the limitations of their memberships. This, among other things, led to a steady move away from developed/developing alignments as the negotiations unfolded and an increasing reliance on regional and issue-specific alignments.

Among the regional groups on the developing side, the Latin American and African groups unquestionably played the leading roles. There was no equivalent to them among the Asian states, and although the Arab states had organized themselves into the Arab League, this organization did not take an interest in the law of the sea until much later. Some of the Arab states were, of course, members of the OAU, but for the others, and for the Asian states, the Group of 77 remained the dominant external alignment.

The Group of 77 largely superceded possible rivals such as the nonaligned states group, which had made the Lusaka Statement in 1970. Other organizations that considered law of the sea questions, like the Asian-African Legal Consultative Committee (AALCC), restricted their activities mostly to discussion and debate, and did not try to play a political role in negotiations at the global level.

The Latin American group did not begin to play an active role until 1970, despite the fact that it possessed a long-standing organization with a formidable record of accomplishment on the law of the sea dating from the previous round of international negotiations. In 1970, the Montevideo and Lima Declarations announced the culmination of the unilateral claims process and its transformation into a coordinated regional movement. The 14 states involved included nearly all the major Latin American coastal states, and their common stand marked a dramatic advance from the relatively isolated three-state grouping of Chile, Ecuador, and Peru. Because it did not include all the Latin American states, this group had some issue-specific characteristics, and should be referred to as the Latin American coastal state group. Its concern was predominantly with the nature and limits of coastal state jurisdiction, and only marginally with deep seabed and other issues beyond that boundary. To solve the problem of internal unity, the group adopted the tactic used so successfully in earlier negotiations on the continental shelf: it created a flexible formula (unilateral determination of maritime boundaries by the coastal state) in order to avoid disputes over particular limits (200 miles) and particular rights (territorial sea versus economic zone).

The Latin American coastal state group had three main strengths, which taken together made it a formidable power. First, it was an almost coherent geographical entity. This characteristic not only facilitated its ability to coordinate policy, but also gave it a very strong regional fallback position if it was unable to get its policy adopted at the global level. The possibilities for such a regional fallback had already been successfully demonstrated by Chile, Ecuador, and Peru in the years since 1952. Second, the group's policy was better suited to the dilemma at hand (fixing a boundary for the international area) than the exploitation formula established in 1958. This put it in an influential position in the SBC negotiations. Third, the group was well placed to exercise influence within the larger Group of 77. Its policy of broad coastal state jurisdiction found considerable sympathy among other coastal developing countries, and as a coherent lobby it was able to some extent to use the Group of 77 as a source of support for the coastal state position.

On the whole, the weaknesses of the group did not detract much from its fundamentally strong position. The fact that it was not completely synonymous with the Latin American group opened it up to regional criticism from countries like Jamaica, Bolivia, and Paraguay that did not benefit from the ambitious coastal state stand. In practice, however, this was not much of a problem. The Latin American group as a whole could nearly always be found on one side of a vote, with at worst one or two states abstaining.[19]

Because its members were regionally concentrated, the Latin American coastal state group was also exposed to the danger of isolation. Its voting strength, though substantial, was not large enough to be decisive in either the Group of 77 or the General Assembly. In addition, the group carried the burden of established opposition by the maritime powers to broad coastal state claims resulting from unilateral actions by many of its members. This danger, though real enough, was largely offset by the group's regional fallback option and by its solid position in both the SBC negotiations and the Group of 77.

By the end of 1970 the group was well launched into a vigorous campaign to promote the strong coastal state position. It was pursuing its diplomatic offensive not only regionally but also within the Group of 77, the SBC, and the General Assembly. It had seized the initiative on the question of a boundary for the international seabed area and was using the double-edged nature of the question to turn the emphasis of the law of the sea discussions away from the international area toward the more immediate issues of coastal state jurisdiction.

At this stage the African group was only just beginning to surface as an actor in seabed politics. Up to the end of 1970, the OAU had not taken any initiatives on the law of the sea, and the African states had therefore concentrated their action within the Group of 77 and the nonaligned state group. Unlike the Latin American group, the African group was relatively new and had no traditions on law of the sea questions. Only four of its members had made exceptional unilateral claims (Guinea, Cameroon, Congo, and Dahomey), and these displayed none of the unity apparent in the Latin American 200-mile claims.

The strengths and weaknesses of the African group were much the same as those of the Group of 77. As a bloc of 41 votes, the group occupied a decisive position in almost any forum where it was active, and like the Latin American group it possessed a relatively strong regional organization, the OAU, within which to coordinate group policy. However, it suffered from a lack of relevant expertise, and from diverse internal interests among its members. Very few African states had coastlines comparable to those of the major Latin American states, and 13 of the 41 OAU members were landlocked. The consensus procedure of the OAU made common policy difficult to formulate, and for this among other reasons the African states had not arrived at a group position by 1970. Thus, as of 1970 the African group existed more as a potential than as a reality in the law of the sea alignments, but its massive voting power made it a potential of great significance. The group awaited only a common position to make it a major actor in the negotiations.

Of the regional groups among the developed states, the Soviet and the western European were the most important. Like the developed state school as a whole, they were weak in voting strength, internally divided by geographical and economic interests, and very strong in economic and technical matters. The Soviet group operated under relatively tight internal discipline, and in combination with its staunch opposition to developing country initiatives, this made it the most visible of all groups. The western European states frequently voted together and possessed institutions within which they could discuss and coordinate policy, but as of 1970 they had made little progress in consciously acting together as a group in the law of the sea negotiations. Both groups were weakened by their devotion of energy to a rearguard defense of maritime state vested interests, and by their consequent lack of coherent new policies taking into account the changed environment of ocean politics. Their economic and technical dominance gave them the capacity to play a major part in formulating new law of the sea policies, but they did not respond to this challenge, choosing instead to use their strength to oppose pressure for change from elsewhere. In this sense, up to 1970 the western European and Soviet groups did not amount to much more than variations in intensity within the overall negative stance of the developed state school.

In comparison with all of these external groups, the issue-specific alignments labored under a number of disadvantages. Since they were largely creations of the law of the sea negotiations, most of them lacked existing institutions and working habits, and consequently had to go through a strenuous and time-consuming process of creation to rise from school to group status. This process frequently involved a considerable amount of political resistance in the sense that membership in such a group required the formation of links across the lines of more familiar regional or developed/developing alignments. Furthermore, such groups were relatively ephemeral in that they were often valid only for the particular negotiation at hand. Consequently no political capital was, or could be, built up within them except in the context of law of the sea issues. This situation was in contrast to that within the more permanent external groups, where support on a law of the sea issue could be used to pay off debts from other old issues or to build up credits for future ones.

Thus, in the context of the seabed and law of the sea negotiations, the long-term advantages of external alignments had to be balanced off against the inappropriateness of such groups in relation to the issue at hand. Correspondingly, the short-term advantages of issue-specific alignments in terms of appropriateness had to be balanced off against both the difficulties of forming and maintaining groups, and the possible costs in terms of other long-range interests.

By the end of 1970, the seabed negotiations had created sufficient pressure for the process of issue-specific alignment formation to begin. This pressure took the form of the increasingly precise definition of areas of dispute described in the previous chapters. The discussion in the SBC and elsewhere revealed and then refined problems such as what boundary to fix for the international seabed area and what relationship seabed exploitation should have to existing production of the same metals. As the discussion progressed, it resulted in the identification of schools, which increased the incentives for states with similar interests to consult and work out common positions. Naturally enough, this process tended to begin within the existing external groups where lines of communication and habits of familiarity made consultation and cooperation easiest. Perhaps the clearest example of this was the emergence of the Latin American coastal state group in 1970, which because of its size quickly dominated the Latin American group as a whole.

The next stage in the formation of an issue-specific group was the establishment of links between the subgroups of different external groups, or between subgroups and individual states where no subgroup had formed. Formation of links at this level marked the creation of an independent issue-specific group that had transcended external group patterns. At any given point in the negotiations it was thus possible to identify: 1) potential issue-specific groups on the basis either of conflict areas in the discussion or of visible schools; 2) emergent issue-specific groups on the basis of evidence indicating formation of issue-specific subgroups within active external groups; and 3) independent issue-specific groups on the basis of visible cooperation on a particular issue among states from different external groups.

By the time the expanded SBC began to meet in 1971, issue-specific groups could be identified on all three of these levels. The shift in the negotiations from the seabed to the law of the sea brought to the surface several disputes that might serve as the focus for opposed sets of issue-specific groups. Sensitivity over the issue of transit through straits offered possibilities for a shipping state group versus a strait state group. Increasing concern over fishing suggested a distant water fishing group in opposition to coastal states seeking to extend their jurisdiction over fishing grounds. Heightened awareness of marine affairs in general suggested possible groups among states with peculiar geographical characteristics that might require special consideration, such as archipelagic states. Actual formation of such groups depended on many factors, not the least of which was whether or not one of the existing external groups adapted its own policies to the issue, and thereby co-opted or obviated the issue-specific group. In a case such as the shipping states, for example, some or all of the developed state groups might adopt the interest as their own and thereby remove the need for a special group.

Emergent groups could be identified among states that exported the metals found in manganese nodules and also among coastal states, in both cases as a response to disputes that had arisen relatively early in the seabed discussions. The influence of metal-exporting states first became evident within the African group in the 1970 Lusaka Statement's reference to control over prices by the seabed authority. Evidence of further progress in this group was provided by Resolution 2750A, calling for a study of the impact of seabed mining on developing country exporters; the sponsors included the four members of the copper exporters' council (CIPEC): Chile, Peru, Zaire, and Zambia. Nearly all of the 13 other sponsors, however, were prominent members of the Group of 77, and were states with little or no direct interest in the metal export question.

This sponsorship pattern seemed to indicate that the Group of 77 might co-opt the exporting state group before it could grow to include developed country exporters like Canada, Australia, and South Africa. Since the issue was well suited to the overall Group of 77 position on closing the economic gap, and since it did to a considerable extent divide along developed/developing lines, the future of this group appeared even in 1970 to be as a subgroup of the Group of 77. Good reason for the exporting states to remain within the congenial bounds of the Group of 77 could be found in the numerical weakness of the group—a weakness that would be maintained even were it to attain its full potential strength. By converting the voting strength of the Group of 77 to its support, it was much more likely to achieve its objectives than it was by acting independently. At this stage, the natural opponents of the exporting states—the importing states and the states interested in seabed mining, in both cases overwhelmingly composed of developed countries—existed only as potential groups.

The coastal state group was by far the most advanced in Latin America, where, as indicated above, it had all but taken over the Latin American regional group. Unilateral actions in Asia and Africa, particularly the 130-mile territorial sea claim of Guinea and the broad fishing zone claims of Ghana, India, Sri Lanka,

Pakistan, and South Korea, indicated sympathy for the coastal state position but had not yet evolved into visible groupings. Much the same was true among the Western developed states, where Canada and Iceland also had extended fishing claims (and pollution control claims in the case of Canada), and where Canada, Iceland, Norway, and Australia were becoming conspicuous as a result of the differences between their attitude toward the negotiations and that of most other Western states. The four had been among the few developed states to sponsor the declaration of principles, and all but Australia were again among the few developed states sponsoring Resolution 2750C. All four also had coastal interests that clearly outweighed their maritime interests.[20]

Perhaps the most striking feature of the emergent coastal state group was the extent to which its potential membership crossed external group lines. The reason for this was to be found in the essentially geographical basis of the group's interest, as opposed to the essentially functional interest underlying potential groups like distant water fishing states and shipping states. For the most part geographical interests were fairly randomly distributed among the various external groups, as in the case of the metal exporters, whereas interests related to the usage of the oceans tended to correlate strongly, though not exclusively, with developed/developing alignments, as in the case of shipping states, metal importers, and distant water fishing states. However, mere possession of coastal state characteristics or geographical characteristics relevant to the law of the sea negotiations (for example straits, archipelagos, or a continental margin) did not necessarily mean that the state concerned would be attracted to a group based on that geographical interest. Membership in such a group implied that the states concerned held similar views as to the significance of their common attribute, and this was patently not the case in situations where other interests intervened. Canada and the United States, for example, possessed roughly comparable coastal state attributes, but in the United States attitudes towards these attributes were conditioned by extensive and opposed maritime interests, while in Canada the largely unencumbered coastal attributes themselves dominated perceptions.

Although not yet fully formed, the coastal state group had several significant strengths in relation to the seabed and law of the sea negotiations. The number of states with unencumbered coastal interests was easily large enough to amount to a potential voting block larger than the African group, and the global distribution of these states meant that the group was unlikely to follow the path of the metal exporters in becoming a subgroup of an existing external group. Furthermore, the group's interest covered a large number of the issues at hand in the law of the sea negotiations, including control over seabed resources on and under the continental margin, coastal fisheries, coastal navigation, and pollution and environmental protection. This possibility of linking together several issues made the coastal state group potentially both more attractive to members and stronger in terms of bargaining leverage in the negotiations. Like the Latin American coastal state group, its members also had strong unilateral options to bargain with if the negotiations did not seem to be taking adequate account of their interests.

Another factor not to be neglected among the strengths of the coastal state group was the power of the historical imperative on which it was riding. The expansion of coastal state awareness and activity has been a theme of this study, and by 1970 the progress of discussion on the international seabed area in the SBC and elsewhere had once again reopened the coastal state question on the international level. The shift in emphasis in the SBC discussions from the international area to the coastal zone can be taken as a milestone in the fortunes of the coastal state group, and the Latin American coastal state group unquestionably took the lead in bringing this about.

The major weakness of the coastal state group, in addition to the difficulties of formation and maintenance common to all issue-specific groups, arose directly out of some of its strengths. This was the problem of reaching agreement on a unified set of policies. For the coastal state group the normal difficulties of arriving at a commonly acceptable policy position were enhanced by the number of issues with which its members were concerned, the different approaches to which they had committed themselves unilaterally (ranging from functional zones to full territorial seas), and the strength of their unilateral options, which applied to political bargaining within the group as well as that between it and other groups at the negotiations. These difficulties, and the consequent problems for the emergent group in achieving independent status and a unified stand regarding coastal state demands, constituted the major barrier to mobilization of the group's full potential strength. Although there was widespread agreement among states with coastal interests that there was a need for substantial extension of coastal state jurisdiction beyond 12 miles, there were substantive disagreements over how the limits of that jurisdiction should be determined and over the degree of coastal state control over activities within those limits. Because of these disagreements, the formation of a multiplicity of groups and schools on individual coastal issues was as likely an outcome as the formation of a single coastal state group.

This weakness in the emergent coastal state group was crucially important for the future of both the group itself and the negotiations as a whole. Since the coastal state interest was one of the major factors in the law of the sea negotiations, it made a big difference whether or not that interest was represented by a coherent policy position. If it was not, then the internal politics of the emergent coastal state group became subject to pressures from noncoastal interests (external groups, maritime, landlocked and shelflocked states). Such a development would not only add further to the difficulties of group formation, but would complicate the negotiation as a whole by elevating the internal conflicts of one particular interest to the status of a global problem.

By 1970 only one independent issue-specific group had formed; that was the landlocked group. The reason why this group had developed further than the others was that it alone had a history stretching back to the previous law of the sea conferences. The landlocked states had conferred together on the question of access to the sea before the 1958 conference, and as soon as the Seabed Committee negotiations opened the way to another conference, this issue brought them together again. The first sign of landlocked state activity in relation to the SBC came with the participation of a substantial group of African and Asian landlocked states in the Lusaka Statement. Their participation did not, however, indicate much about policy, and it was not until Resolution 2750B that the existence of an independent group became manifest. The resolution called for a study of the problems of landlocked states in relation to both the international seabed regime and free access. It was sponsored by 12 landlocked states, of which eight were African, two Latin American, one Asian, and one Western (Austria).[21] This list of sponsors included members of all external groups except the Soviet and Arab groups (there being no Arab landlocked states), thereby placing the landlocked group further ahead in terms of development than its most obvious opponent, the coastal state group.

In addition to its advantage of relatively early formation, the landlocked group had a number of other strengths. As a voting bloc it had a potential membership of 27 states (eight developed, and 19 developing of which 13 were African)—a considerable but not decisive weight. Its voting strength became much more impressive when considered in the light of potential allies. The shelflocked states, of which there were 20 (nine developed, and 11 developing of which six were Arab), shared many though not all of the problems faced by the landlocked states. Taken together the two sets of states amounted to about one-third of the votes in the international community, and were therefore a power to be reckoned with in any attempt to mobilize winning majorities. A second possible source of allies for the landlocked states was among the maritime powers, which would be likely to share the objective (although for different reasons) of restraining coastal state expansion.

Another strength of the landlocked states was that, like most of the developing countries, they had little to lose and potentially quite a lot to gain from the negotiations. Their peculiar geographical circumstance meant that policy for the group was a relatively straightforward matter of pressing for equality of status, freedom of access, and maximum sharing of ocean wealth within the international community. Because of their voting strength, the landlocked states could not be ignored in an international negotiation, and were therefore in a reasonably strong position to bargain for concessions in return for voting support. The same argument applied to the position of the landlocked states in the regional groups and the Group of 77. Although their strength varied from very weak in the Latin American group (two) to very strong in the African group (13), they were generally well placed to advance their interests both within and outside the external groups.

Against these strengths, the landlocked group suffered from three weaknesses. First, nearly all of the 19 developing landlocked countries were among the poorest and least technically capable states in the world. Their ability to participate either in the negotiations or in ocean activity was therefore limited. Second, and partly because of this, many of the landlocked countries were vulnerable to pressure from neighbors and regional groups. Landlocked states were frequently the weakest members of a regional group, and since any implementation of rights for the landlocked, especially transit rights, would require local and regional cooperation, there were disincentives for them to move too much into conflict with majority regional interests. Third, the landlocked states by definition had no unilateral options whatever in relation to any of the law of the sea issues, and therefore had no fallback position should the negotiations turn against their interests. This made them wholly dependent on the success of the negotiations to protect their share of the common heritage from unilateral expropriation by coastal states.

By the beginning of 1971, then, the major alignments in seabed politics were already clearly developed. Progress in the negotiations between 1968 and 1970 had stimulated the formation of several important new groups, and had made clear which existing external groups were going to play key roles in the negotiations. External groups had unquestionably dominated alignments during the first three years of negotiations, but by 1971 issue-specific groups were beginning to assume greater weight. Only a few alignments related to seabed issues alone, while many had interests that included seabed and other issues, both law of the sea and non-law of the sea. Some alignments had little or nothing to do with seabed issues. This mixed nature of the major alignments meant that seabed politics was inextricably tangled up with other issues, again both law of the sea and non-law of the sea. Bargaining positions therefore had to be seen not only in the light of the issue at hand but also in the light of other group interests and of the overall effort to achieve a package deal on the law of the sea by trading concessions among as well as within individual issues.

In spite of the mixed nature of the major alignments, seabed politics still maintained a substantially separate existence from other aspects of ocean politics. This occurred partly because of the physical differences between the seabed and the water column, partly because of the separation between the two established in both the 1958 Continental Shelf Convention and the 1970 declaration of principles, and partly because discussion of the seabed issue had started independently from and earlier than the more general law of the sea debate. Although there were links between seabed and water column issues (such as the idea of an ocean space regime and machinery beyond the limits of national jurisdiction, and the idea of tying the seabed and the superjacent waters together in a fixed distance economic zone or territorial sea), these links were not as strong as those among the more numerous and complex issues centered in the water column. It was much easier to discuss the seabed issue on its own merits than it was, for example, to discuss the breadth of the territorial sea in separation from a host of other concerns like navigation, fishing, and environmental protection.

Inasmuch as seabed politics could still be treated as a separate area of the negotiations, it was possible to divide it into two sections. The older of the two sections comprised the effort to devise an international regime and machinery for the seabed area beyond the limits of national jurisdiction. Since this question had been the dominant focus of negotiations between 1968 and 1970, positions on it were relatively clearly developed. By 1971, however, it was declining in importance beside both other seabed and other law of the sea issues. The primary alignment here, as illustrated in Table 6.3, was the opposition of the developed and developing countries. The metal-exporting states looked likely to become a subgroup of the Group of 77, but there were no indications that this fundamental pattern would undergo changes other than the elaboration of further issue-specific subgroups, like the metal importing states, within the dominant external group alignment.

The younger of the two sections comprised the effort to fix a boundary for the limits of national jurisdiction over the seabed and to determine the extent of coastal state control within that boundary.

TABLE 6.3

Major Positions in the Debate on International Machinery, 1968-70

Opposed to Establishing International Machinery[a]	Favor Establishing Weak Machinery Along Registry/ Weak Licensing Lines[b]	Favor Establishing Strong Machinery Along Exploiting/ Strong Licensing Lines[c]	
		(LATIN AMERICAN)	(AFRICAN)
Bulgaria	United States	Brazil	Cameroon
Czechoslovakia	United Kingdom	Mexico	Kenya
Hungary	France	Chile	Mauritania
Mongolia	Japan	Colombia	Nigeria
Poland	Italy	Ecuador	Tanzania
Romania	Australia	El Salvador	Chad
Soviet Union	Canada	Guatemala	Zaire
		Guyana	Gabon
		Honduras	Ghana
		Nicaragua	Ivory Coast
		Peru	Liberia
		Jamaica	Malagasy Republic
		Trinidad and	Mali
		Tobago	Mauritius
		Barbados	Niger
			Senegal
		(ASIAN)	Swaziland
		India	Togo
		Sri Lanka	
		Cyprus	(ARAB)
		Pakistan	Kuwait
		Philippines	Libya
		Thailand	Sudan
		Indonesia	Lebanon
		Malaysia	Morocco
		Singapore	
		Turkey	
		Maldives	
		(OTHER)	
		Yugoslavia	

[a]Statements in the SBC or vote against Resolution 2467C (first study on international machinery).

[b]Statements in the SBC.

[c]Statements in the SBC or sponsor of Resolution 2574C (second study on international machinery).

TABLE 6.4

National Limits Positions in the Debate on the International Area, 1968-70

	Geological Margin	200-Meter Isobath	500-Meter/200-Mile Depth/Distance Formula	200 Miles	Unilateral Determination by Coastal State
States advocating	Canada	United States	Norway	Malta*	
	Italy		Iceland	China*	China
	Australia		United Kingdom*	Peru	Peru
	United States*			Argentina	Argentina
	Argentina			Brazil	Brazil
				Chile	Chile
				Ecuador	Ecuador
				El Salvador	El Salvador
				Nicaragua	Nicaragua
				Panama	Panama
				Uruguay	Uruguay
					Colombia
					Dominican Republic
					Guatemala
					Honduras
					Mexico
States opposing	Tanzania	United Kingdom	United States	United States	Malta
	Liberia	Australia			United States*
	Sri Lanka				Soviet Union*
					Tanzania*

*Tentative commitment to position

Sources: Statements made in the SBC, Montevideo and Lima Declarations, and statements made outside the SBC (for the United States and China).

In a sense this was, in fact, the older of the two sections since it had been dealt with at the earlier Law of the Sea Conference, but in terms of the negotiating round starting in 1968, it was the younger and less clearly developed of the two. As we have seen, this question had arisen out of the international regime debate, and had been the lever used to shift the negotiations toward a reopening of the whole law of the sea. Partly because of this use, it was more encumbered by associated water column issues like the width of the territorial sea or economic zone than was the international regime question. The primary alignment here looked like being coastal states versus landlocked and shelflocked states, but as Table 6.4 illustrates, only the coastal states led by the Latin Americans had as yet declared themselves. Such opposition as was visible to the coastal states came mainly from maritime states and was more on the basis of water column than of seabed concerns. As Table 6.4 also shows, however, the coastal states themselves were arrayed in several small groups and schools, and were still far from a unified position.

NOTES

1. Appendix D of Northcutt Ely's statement in hearings before the Special Subcommittee on the Outer Continental Shelf of the Committee on Interior and Insular Affairs, U.S. Senate, 91st Congress, December 1969-September 1970, Washington: USGPO, 1970. Vol. 1 (of 3), pp. 61-65, is a useful source on continental shelf claims up to 1967. He lists, inter alia, 11 countries and five colonies not listed in Table 6.3 that have granted offshore concessions in the absence of general legislation. Taken together with the 100 other states listed in Table 6.2, this would give a total of 111 coastal states and territories recognizing the principle of coastal state jurisdiction over adjacent offshore mineral resources. On national claims generally, see U.S. Department of State, Limits in the Seas, "National Claims to Maritime Jurisdiction," Bureau of Intelligence and Research, Geographer 36, second rev. (April 1974).

2. Choon-Ho Park, Continental Shelf Issues in the Yellow Sea and the East China Sea, Law of the Sea Institute Occasional Paper no. 15, (Kingston: University of Rhode Island, September 1972), pp. 3-5.

3. S. Oda, "The Delimitation of the Continental Shelf in South East Asia and the Far East," Ocean Management 1, no. 4 (December 1973): 335-37.

4. Hungdah Chiu and Choon-Ho Park, "Legal Status of the Paracel and Spratley Islands," Ocean Development and International Law 3, no. 1 (1975): 1-28.

5. Park, op. cit., pp. 21-6.

6. F. J. Gardner, "Huge Offshore Oil Growth Set for 1970's," Oil and Gas Journal 68, no. 11 (March 16, 1974): 127.

7. Vincent E. McKelvey and Frank F. Wang, World Subsea Mineral Resources: Preliminary Maps, U.S. Geological Survey, Department of the Interior, 1969, p. 8.

8. World Oil 171, no. 1 (July 1970): 84.

9. John A. Knauss, "Factors Influencing a U.S. Position in a Future Law of the Sea Conference," LSI Occasional Paper no. 10 (Kingston: University of Rhode Island, April 1971), p. 12.

10. Gardner, op. cit.

11. L. Sheffer, "Where Those Offshore Rigs Are," in ibid., pp. 152-88.

12. World Oil 172, no. 1 (January 1971): 74.

13. Albert E. Utton, "A Survey of National Laws on the Control of Pollution from Oil and Gas Operations on the Continental Shelf," Columbia Journal of Transnational Law 9, no. 2 (1970): 331.

14. D. J. Walmsley, "Oil Pollution Problems Arising Out of the Exploitation of the Continental Shelf: The Santa Barbara Disaster," The San Diego Law Review 9, no. 3 (May 1972): 516.

15. Reprinted in U.S. Senate, Mineral Resources of the Deep Seabed, 93rd Congress, first session on S.1134, pp. 387-97.

16. Jerzy Sztucki (ed.), Symposium on the International Regime of the Seabed (Rome: Academia Nazionale dei Lincei, 1970), p. 121.

17. "Most active" defined on the basis of sponsorship of proposals. The other seven active developing countries were Kenya, Tanzania, Malagasy Republic, Thailand, Libya, Kuwait, and Sudan.

18. E. Miles, "The Dynamics of Global Ocean Politics," in Douglas Johnston (ed.), Marine Policy and the Coastal Community: Studies in the Social Sciences (London: Croom Helm, 1976) borrows from Olav Knudsen, The Politics of International Shipping (Lexington: D. C. Heath, 1973), pp. 76-83, 96-100, the concepts "pure" and "contaminated" in relation to the penetration of ocean issues by external issues. These concepts can be applied in a complementary way to alignments as well as to issues.

19. For example, in a 1970 vote in the First Committee of the General Assembly on a proposal to charge the SBC with the task of delimiting the area (A/C.L/L.564), all but two of the Latin American states (Mexico and Cuba abstained) were to be found opposing. This group vote, which included the two landlocked states, indicated the extent to which the Latin American states worked together even when it was not self-evidently in their interest to do so. A/C.1/PV.1800, p. 15.

20. For a detailed analysis of Canada on this point, see B. G. Buzan and B. Johnson, Canada at the Third UN Law of the Sea Conference: Policy, Role and Prospects, LSI Occasional Paper (Kingston: University of Rhode Island, 1975).

21. Miles, op. cit., identifies Austria as the leader in mobilizing this group.

CHAPTER

7

SEABED POLITICS AND THE LAW OF THE SEA, 1971-73, PART I: ELABORATING A REGIME FOR THE INTERNATIONAL SEABED AREA

The transformation of the Seabed Committee by Resolution 2750C resulted in a much more extensive and complex agenda covering all law of the sea issues. Instead of being a moderately sized committee dealing only with the problems raised by Pardo's proposal on the deep seabed, as it had been from 1968 through 1970, the SBC was now a large preparatory committee charged with laying the foundations for a Third UN Conference on the Law of the Sea. In this context, five more states were added by Resolution 2881 of December 21, 1971—China, Fiji, Finland, Nicaragua, and Zambia—thus bringing the formal membership to 91. This transformation of the SBC necessitates a change of approach in the present study, since not all aspects of the committee's work from 1971 through 1973 were relevant to seabed politics. The retention of the name Committee on the Peaceful Uses of the Sea-Bed and the Ocean Floor Beyond the Limits of National Jurisdiction, unchanged from 1969 through 1973, has been a source of some confusion in this regard for those unfamiliar with the subject.

With its expanded mandate, the enlarged SBC required different subcommittees from those that had served its predecessor. It set up three completely new subcommittees of the whole in a pattern that was to remain unchanged through the lifetime of both the committee and the following conference. Subcommittee I (hereafter SC.I) was instructed to prepare draft treaty articles embodying the international regime and machinery for the seabed beyond the limits of national jurisdiction. It was to take the declaration of principles (Resolution 2749) as its starting point, and so can be seen as the direct successor of the old SBC. SC.II was given the formidable task of preparing "a comprehensive list of subjects and issues relating to the law of the sea, including those concerning the regimes of the high seas, the continental shelf, the territorial sea (including the question of its breadth and the question of international straits) and contiguous zone, fishing and conservation of the living resources of the high seas (including the question of preferential rights of coastal States) and to prepare draft treaty articles thereon." SC.III was charged with preparing draft articles on the preservation of the marine environment, including prevention of pollution, and on the question of scientific research.[1]

The committee reserved for its plenary body the question of defining a boundary between national jurisdiction and the international seabed area and the question of peaceful uses of the area, on the grounds that these were issues where controversy was already well established. Since the boundary question was vitally important to the work of both SC.I and SC.II, the subcommittees were allowed to discuss it in relation to the appropriate items on their agendas. In practice, discussion of the boundary issue tended to concentrate in SC.II, where it became tied in with the question of the limits of coastal state jurisdiction in the water column. In terms of seabed politics, then, our attention must focus on all the work of SC.I; the continental shelf, coastal economic zone and territorial sea aspects of the work in SC.II; and the discussion of the boundary in the plenary meetings of the committee. Given the firm direction set down in the new mandate, and since the process of work in the expanded SBC was much the same as it had been in the old SBC, it is not necessary to trace the annual progress of the committee as was done in previous chapters. Instead, the three-year period from 1971 through 1973 will be treated in this and the following chapter as a single unit, and analysis will center on the twin themes of defining the boundary between national and international jurisdiction on the seabed, and elaborating a regime for the international seabed area.

NODULE MINING: ECONOMIC POLITICS

The effort to elaborate an international regime and machinery for the seabed beyond the limits of national jurisdiction took place against the background of ongoing developments in the nodule mining industry. These developments included progress made by private companies in creating appropriate organizations and technology to undertake deep sea mining, continuation of the economic aspects of the great debate, particularly the attempt to assess how much revenue would be generated by nodule mining and how much impact such mining would have on existing metal markets, and political maneuvering in the United States aimed at promoting national jurisdiction over deep sea mining.

Among the mining companies, the pace of activity quickened both in regard to the number of corporations interested in deep sea mining and the level of their research and development. U.S. companies maintained their early lead, but were joined in the technology race by others from Japan, West Germany, France, Canada, and Australia. The main contenders, and their activities, were as follows.[2]

United States

Deepsea Ventures (Tenneco) was among the first comers to the field, and by 1973 had undertaken over 30 research cruises in the Pacific resulting in identification of several potential mine sites. The company had conducted a number of tests with prototype nodule processing pilot plants, had tested an air-lift hydraulic dredge to 2,400 foot depths, and had spent some 20 million dollars.

Kennecott Copper Corporation was also an early comer, and like Deepsea Ventures had conducted extensive survey cruises resulting in identification of potential mine sites. Kennecott had tested pilot processing plants, but despite a substantial research program had not tested a mining system.

Ocean Resources worked mostly through international consortiums, and had established close links with the Sumitomo Group in Japan. The company took part in the second and third tests of a continuous line bucket mining system (1970 and 1972) which were supported by an international consortium of 32 interested corporations.

Summa Corporation (Hughes) climbed on the nodule bandwagon in 1969, and concentrated most of its effort on developing ocean mining technology. In mid-1971 the company began construction of what was thought at the time to be a 35,000-ton ocean mining ship and a dredge-head launching barge, and these were thought to be nearly ready to begin testing by the end of 1973. Summa had the advantage of internal financing, and consequently was not affected by the legal uncertainty surrounding nodule mining that made financing a problem for many other companies. Estimates of Summa's outlay ranged up to 250 million dollars,[3] and other U.S. companies like Lockheed and Global Marine were heavily involved in Summa's program.

Japan

The Sumitomo Group, with the support of the Ministry of Trade and Industry, took up an interest in nodule mining in 1968. Tests of a CLB system began in 1968, and were continued with support of the consortium mentioned above. Of the 32 companies participating in the consortium seven were American, one German, one Australian, two French, and six Canadian. The 1972 test was on a larger scale than the previous two and apparently ran into some difficulties, but the group remained optimistic about the CLB system and hoped to have one in operation by 1977.

The Deep Ocean Mining Association (DOMA) was a late entry into the nodule mining business. It represented a consortium of 27 Japanese companies, and appeared to be starting from scratch at its formation in April 1973.

West Germany

AMR was a joint venture by three West German companies supported by government funds. It succeeded earlier joint ventures, also government supported, that had undertaken research cruises in collaboration with Deepsea Ventures in 1970 and 1971. The group was beginning work on processing and mining systems, and had been approached by Kennecott to set up a multinational joint venture.

France

CNEXO/Societe le Nickel was a research partnership based on earlier work done by CNEXO alone. CNEXO undertook prospecting cruises around French Polynesia during 1970-71, and by 1972 had completed setting up a Pacific Oceanographic Center. Both companies participated in the CLB testing consortium.

Canada

INCO was one of the six Canadian companies participating in the CLB consortium, and its own interest in nodules went back to the early 1960s. The company undertook several exploration cruises, but most of its work appeared to be concentrated on research rather than on actually building mining and processing systems itself. Noranda Mines also maintained an interest in the nodules.

As a result of this intense commercial activity it was becoming clear to observers all through the period 1971-73 that the technological problems of nodule mining were very close to solution, and that if events were allowed to run their course commercial nodule mining would be likely before the end of the decade. The enthusiasm of the mining companies indicated a favorable evaluation on their part of the economic prospects for nodule mining, and their activity provided a continuous flow of data to the debate on likely revenues from, and likely impacts of, such an industry. Because there was still no operational data from a nodule mining operation, however, this debate continued to be plagued by the same difficulties that had affected it in the late 1960s. By 1971 the main activity in the debate had shifted to the SBC and the General Assembly, and it will be discussed in that context below.

It is interesting to note also that commercial interest in nodule mining was not confined to countries with importing interests in the four metals. Although the United States, Japan, West Germany, and France were all large importers, Canada's interest was that of a large exporter. Italy, Britain, and Sweden showed little interest in nodule mining despite being large importers, and the Soviet Union, a large exporter, went only so far as to set up an International Coordinating Center of Marine Exploration among the COMECON countries to undertake exploration of the mineral resources of the seabed. The Soviets showed no visible interest in nodule mining technology for themselves.

In the United States, progress toward an operational nodule mining industry generated an important political movement in support of national jurisdiction over exploitation of the deep seabed. The prime focus of this movement was the Deep Seabed Hard Minerals Resources Bill (also known as the Hard Minerals Bill) first introduced in Congress by Senator Lee Metcalf in November 1971. The Hard Minerals Bill (known as S.2801/HR.13904 during the 92nd Congress and as S.1134/HR.9 during the 93rd) strongly reflected the position of the U.S. mining industry. Its purpose was to "provide the Secretary of the Interior with authority to promote the conservation and orderly development of the hard mineral resources of the deep seabed, pending adoption of an international regime therefor."[4] To this end the bill provided for, among other things:

(1) Nodule mining license blocks 40,000 square kilometers in extent;

(2) Free rights to scientific research and prospecting on the seabed outside licensed areas;

(3) No requirement for license holders to divulge proprietary information;

(4) An international registry organization restricted in function to record keeping in respect of licensing;

(5) Licensing on a first-in-line basis subject only to applicants paying a small license fee ($5,000) and meeting technical and financial requirements to be set by the Secretary of the Interior. Licenses could be revoked only if the license holder failed to remedy his violation within a reasonable time period after being informed of it by the Secretary of the Interior.

(6) Establishment of an escrow fund based on a percentage of all license and tax revenues accruing to the government from activities under the act. U.S. contributions to this fund would be at the same level as the contributions of other developed states that accepted the legislation, and the revenues would be used as directed by Congress to assist developing countries that also accepted the legislation.

(7) Protection for miners against any interference by others, whether lawful or not, and against any disadvantages resulting from acceptance by the United States of an international regime for the area. Protection against an international regime would include guarantee of the right of the miner to continue to work his license block for the term of the license and reimbursement for any loss of investment or increase in costs incurred by the miner as a result of any new regime.

Full understanding of the Hard Minerals Bill would require a detailed examination of political bargaining in the internal politics of the United States—a task beyond the scope of the present study. Suffice it to say that the hard minerals industry, as represented by the American Mining Congress, played a major role in drafting the bill and was the strongest lobby behind it. Their position was probably best stated by Senator Metcalf, who favored using unilateral action to force the development of international law (along Canadian and Latin American lines): "We are way ahead of a lot of people, and we may lose that lead and our resources if we just sit around here and wait for the State Department to drag its feet while trying to negotiate some international treaty."[5]

The industry felt that the legal uncertainties of nodule mining were inhibiting its ability to finance commercial operations, and that consequently the U.S. lead in technology was being eroded by developments in Japan and West Germany. They were careful to argue for the Hard Minerals Bill as an interim measure, and discounted the accusation that it was unilateral by arguing that it did not affect the rights of any other nation to use the seabed in accordance with existing international law.[6] They also argued that passage of the bill would have a stimulating and beneficial effect on the international negotiations at the United Nations.[7] Their aim, as the terms of the bill made clear, was to institute a legal regime for deep ocean mining that would be favorable to the mining industry in terms of guaranteed security of rights, maximum freedom of activity, and minimum obligations. Even more important, however, was the fact that the bill would strongly tend to fix the U.S. position on an international seabed regime by requiring that the United States not accept any regime that would deprive the miner of his established rights under the national legislation.

The sustained presence of the Hard Minerals Bill before Congress created pressure in two directions: first on the State Department and the executive within the United States, second on the law of the sea negotiations at the United Nations. Internally, the situation was complicated by the fact that the bill could easily be seen as conforming to the guidelines set down by President Nixon in his May 1970 statement on seabed policy. In that statement the President had said:

> I call on other nations to join the United States in an interim policy. I suggest that all permits for exploration and exploitation of the seabeds beyond 200 meters be issued subject to the international regime to be agreed upon. The regime should accordingly include due protection for the integrity of investments made in the interim period. A substantial portion of the revenues derived by a state from exploitation beyond 200 meters during this interim period should be turned over to an appropriate international development agency for assistance to developing countries. I would plan to seek appropriate congressional action to make such funds available as soon as a sufficient number of other states also indicate their willingness to join this interim policy.

Furthermore, the State Department had replied as follows to an inquiry from Senator Metcalf regarding United States policy toward the moratorium resolution:

> The Department does not anticipate any efforts to discourage U.S. nationals from continuing with their current exploration plans. In the event that U.S. nationals should desire to engage in commercial exploitation prior to the establishment of an internationally agreed regime, we would seek to assure that their activities are conducted in accordance with relevant principles of international law, including the freedom of the sea, and that the integrity of their invest-

ment receives due protection in any subsequent international agreement.[8]

Despite the apparent sympathy for an interim regime of the type proposed in the Hard Minerals Bill, the executive refused to take a firm position on the bill for more than a year after it was first introduced, and eventually decided early in 1973 to oppose its enactment. The main reason for this appears to have been realization that enactment of the bill would encounter severe opposition internationally. This would come especially from the developing countries, both generally because of the bill's contravention of the common heritage principle as the developing countries interpreted it, and particularly because of the bill's provision for a free hand for mining companies, a very weak international machinery, and a probably very small international fund (since both licensing fees and taxes under the proposal would be relatively small). The executive feared that such opposition might endanger the package law of the sea negotiation to which the United States had committed itself in 1970.[9] The delay in taking a position on the bill provided useful leverage for the United States in the negotiations without appearing to involve the executive directly in a movement toward unilateral action. When the decision to oppose the bill was finally taken, it was justified on the grounds of the threat posed to the international negotiations because others would see it as unilateralism of the type consistently opposed by the United States. The executive clearly could not resolve the dilemma that any interim legislation could also serve as an alternative to an international agreement. Thus while maintaining its support of interim activity as outlined in the Nixon statement, it rejected the Hard Minerals Bill and renounced any other interim legislation in the period before or during the scheduled Law of the Sea Conference (1974-75) so long as a successful outcome to the negotiations appeared probable by 1975.[10]

While thus emphasizing its commitment to, and optimism about, the international negotiations on the law of the sea, the executive also kept open two important options based on national legislation. First, it emphasized the need to design appropriate legislation for the period between signature of a treaty and its entry into force. Second, it maintained the right to adopt alternative legislation in the event that the Law of the Sea Conference did not produce an acceptable treaty within an acceptable time limit.[11] It made this latter option quite explicit in the Seabed Committee debates while at the same time affirming its opposition to the Hard Minerals Bill then before Congress.[12] In this way, the executive was able to have the best of both worlds. Not only could it be seen to be opposing unilateralism within its own country and supporting the international negotiating process, but also it could take advantage of the pressure created by a unilateral threat to serve its own interests in the negotiations—particularly its interest in achieving a workable agreement at the earliest possible time.

The impact of the Hard Minerals Bill on the law of the sea negotiations at the United Nations was mostly absorbed by the debate on the nature of machinery for the seabed regime. The bill was a relatively extreme example of the case for weak international machinery, and since it came from the state most advanced in nodule mining capability its effect was to heighten tensions in the dispute between those favoring licensing machinery and those favoring operating machinery. It also aroused the fears of the metal-exporting states because of the prospect it created of unrestricted production from the seabed competing with established land production. In these disputes the bill served as a convenient "worst possible model" for the developing countries by raising a concrete possibility of the kind of technological imperialism that had originally inspired Pardo to make his proposal.

The major response to the bill came in March 1972, when the representative of Kuwait introduced a draft decision (138/L.11) on the subject in the SBC. The committee decided to defer consideration of L.11 until its second session, and when the item was reintroduced (as L.11 rev.1) in August, it was sponsored by 13 developing countries, including China, and supported by many of the leading states in the Group of 77.[13] The draft decision invoked the moratorium resolution and the declaration of principles, expressed grave concern over the evidence of "operational activities in the area" by "a number of States, organizations and consortia", and called upon

> all States engaged in activities in the sea-bed area, beyond the limits of national jurisdiction, in conformity with the provisions of the two resolutions cited above, to cease and desist from all activities aiming at commercial exploitation in the sea-bed area and to refrain from engaging directly or through their nationals in any operations aimed at the exploitation of the area before the establishment of the international regime.

It went on to reaffirm that

> prior to the establishment of the international regime, no claims on any part of the area or its resources, based on past, present or future activities, will be recognized.[14]

The document precipitated a heated discussion, and revealed that the passions surrounding the original moratorium resolution in 1969 had not diminished during the intervening three years. Chile led the charge for the developing countries by denouncing the Hard Minerals Bill as a violation of the moratorium resolution and a threat to the common heritage of mankind. The Chilean delegate pointed to the danger of unrestricted exploitation for metal-exporting states (including his own), and urged the necessity for production controls. This resulted in an affray between Chile on the one hand, arguing that nodule mining by the developed states was imminent, and the United States, Britain, France, and Japan on the other, arguing that it was relatively remote. The United States declared that it was not prepared to accept a "monopoly by an international operating agency over deep sea exploitation," and argued that international law did not forbid exploitation, that no commercial operations were active anyway, that the moratorium resolution was disputed, that a revival of the moratorium argument would prove divisive and might delay development of technology, that only a total moratorium including all types of coastal state claims would be acceptable, and that the best solution was to work with all possible speed on creating a workable international regime.[15]

The seriousness of the clash and the prospect of its wrecking the negotiations seem to have caused both sides to back away rather than to press the confrontation. The developing countries did not try to revive the moratorium issue in the 1972 meetings of the General Assembly, and did not stress it in the 1973 SBC sessions. For its part, the United States came out against the Hard Minerals Bill in 1973, and stressed the importance of the conciliatory action of the developing countries in not reviving the moratorium issue.[16]

This whole incident illustrated the limitations of the developing countries' voting power when arrayed against the economic and technical strength of the developed states. The steady development by the United States of a unilateral option to undertake nodule mining suffered at most only a minor delay as a result of the 1969 moratorium resolution. Furthermore, on the basis of its progress in deep ocean mining technology and national legislation, the United States in 1972 was able to meet and counter the attempt by the developing countries to reapply the moratorium by raising its own much more solidly based threat of unilateral action.

The incident exemplified the delaying effect that the Hard Minerals Bill had on the international negotiations because of the way in which it precipitated political confrontations. This was in contrast to the claims of some of the bill's supporters that it acted as a spur to the negotiations. Their view ignored the numerous political, structural, and conceptual difficulties that caused the negotiations to be slow, and assumed that slowness was caused by some delegations deliberately dragging their feet. By creating an inexorable momentum toward an operational nodule mining industry, the mining companies certainly exerted formidable pressure on the international negotiations: they confronted all delegates with a fait accompli. But because of the political structure and content of the negotiations, this resulted much more in a threat to the continuation of the negotiations than in creating a sense of urgency about bringing them to a rapid conclusion. Equally important, however, the actions of the companies gave the advanced technology state delegations a powerful lever with which to counterbalance the massive voting weight of the Group of 77 in SC.I.

In the SBC, the continuous flow of reports from the Secretary-General, mostly requested by developing countries, served as a useful source of information on developments in the nodule mining industry. These reports not only provided regular updates on progress in technology, but also undertook analysis of the economic implications of nodule mining and of other aspects of the international regime, such as the boundary of the area and the ways of sharing benefits from exploitation. Three of these reports, A/AC.138/36, 73, and 90, can be considered together since the latter two were simply annual updates of 138/36. They were all compiled in response to Resolution 2750A (December 1970). Their subject was developments in the nodule mining industry and the possible impact of seabed mineral production from the international area on world markets, with special reference to the problems of developing countries.

In 138/36, the Secretary-General injected into the SBC the first large dose of information about the nodule mining industry. The report outlines the uses of the four nodule metals, identified the major exporters and importers, and pointed out the problem of the metals not occurring in anything like the same proportion as market demand for them.[17] On the economic side, the Secretary-General argued that the impact of nodule mining on developing country metal exporters would not be very serious. In the case of nickel, rapidly rising demand would offset the effect of new production, and any drop in price that did occur would largely fall on the shoulders of the three developed states that accounted for nearly 90 percent of world production. These were Canada, France (New Caledonia), and the Soviet Union. Copper was also protected by rapidly rising demand, and was further removed from threat by the fact that nickel and copper would in all probability have to be exploited together in order to make nodule mining profitable. Since world demand for copper was ten times as great as that for nickel, and since both metals occurred in nodules in approximately the same proportions, nodules would account for only a small percentage of the copper market. In the case of manganese, the market seemed safe because of the high cost of extracting the metal from nodules and the relatively small demand for the pure manganese that would result from such a process (most manganese is used in an impure ore form preferred by the steel industry). Only cobalt seemed likely to suffer a drop in price, but since the world market for the metal was small, and since the possibility of substituting cobalt for nickel prevented the price from falling below that of nickel, the impact did not appear likely to be severe. Since one country (Zaire) dominated the export market the problem would not be widespread even if it did arise. The Secretary-General did not consider that forseeable production of oil from the international area would have any impact on existing exporters.

This report was in the nature of a preliminary assessment rather than a definitive statement. It suffered from the same problem that afflicted the great debate in the United States, namely that it was based largely on speculative economic data. As a result, its conclusions were completely open to criticism based on other speculative data, and it had little effect in changing the attitudes of those states with vested interests in the subject. The developing countries tended to criticize the report for underestimating the possible impact on exporting states, while the developed states accepted the Secretary-General's conclusions as just about right. Everyone accepted the need for further study, and some of the developing countries urged a greater role for UNCTAD in preparing subsequent studies.

In his 1972 report on the subject (138/73) the Secretary-General reviewed developments in the nodule mining industry and made a further attempt to come to grips with the question of economic implications. By drawing on a somewhat wider range of nodule mining scenarios, and with the assistance of UNCTAD, the report managed to present a gloomier outlook for the developing metal-exporting states than its predecessor had done. It contained proposals for alleviating the threat to such countries either by compensation or by prevention, and was noticeably favorable to the latter approach, involving production and market controls. The result was a predictable reversal of positions, with the developing countries tending to support the report and the developed states tending to be critical of it. Debate on 138/73 took place in the inflamed atmosphere of the confrontation discussed above over the activities of mining companies and the possibility of a renewed moratorium resolution. Partly as a result of that confrontation and partly as a result of a lack of any new data, the Secretary-General made no further attempts during the life of the SBC to expand on the economic implications of deep seabed mining. His

1973 report (138/90) was confined entirely to a review of developments in the nodule mining industry.

Two other reports of the Secretary-General are also of interest here as inputs to the SBC in the period 1971-73. The first, 138/38, was presented to the committee in 1971 and contained a study of "possible methods and criteria for the sharing by the international community of proceeds and other benefits derived from the exploitation of the resources of the area beyond the limits of national jurisdiction." It surveyed possible financial and nonfinancial benefits and ways of distributing them, but was forced to conclude that given all the unknowns of the seabed regime and machinery not much could be said definitively on the subject.[18] In subsequent years, this aspect of the international regime increasingly took second place to concern over how to alleviate possible adverse effects from seabed exploitation, and consequently was not much discussed.

The second report, 138/87, was a result of Resolutions 3029B and C (December 1972), and was the subject of a controversy that will be considered in the next chapter. It contained a study of the "economic significance, in terms of seabed mineral resources, of the various limits proposed for national jurisdiction." The study examined four possible limits (40 miles, 200 meters, 3,000 meters, and 200 miles) to determine what seabed resources would accrue to the coastal states and to the international regime in each case. The conclusion was unequivocable. The two narrower limits would leave all mine-grade manganese nodules in the international area, would include between 30 and 40 percent of ultimate oil resources (with some near-term prospects), and would also include prospects such as the metalliferous muds in the Red Sea. The two wider limits, by contrast, would leave a maximum of only 13 percent of ultimate oil resources within the international area (none with near- or medium-term prospects), would place some nodules within the jurisdiction of volcanic islands in the Pacific (in the case of 200 miles only), and would leave all other mineral resources to the coastal states.

The report made starkly clear the nature of the choice between national and international jurisdiction. Narrow limits meant that the international regime would receive significant revenues from oil, and therefore that the common heritage of mankind would almost certainly produce sufficient benefits to allow for substantial aid or redistribution to the developing countries. Wide limits meant that the international regime would have to rely exclusively on manganese nodules for its revenues, and therefore that the common heritage would be of uncertain, and possibly quite limited, value to the developing countries. This information came too late to have an impact on the debate (the general conclusions were not new, but were merely stated in precise terms) because the trend toward wide coastal state limits on the seabed was already firmly established by the time the report was submitted.

Like the other Secretary-General's reports on seabed matters in this period, then, 138/87 served a useful education function, but had little effect on the course of the debate. The positions, attitudes, and alignments that had taken shape in the early years of the SBC remained unmoved by inputs of information, and attention concentrated instead on the political implications of the numerous proposals outlining alternative regimes and machineries that were being put forward by states.

ELABORATING A REGIME: PROPOSALS AND POSITIONS IN THE SBC

Between August 1970 and October 1971, 11 states or groups of states submitted proposals on the international regime and machinery to the Seabed Committee. These proposals varied from comprehensive drafts worked out in great detail to short statements bearing on only a few aspects of the subject. All of them owed some debt to the two studies by the Secretary-General made in 1969 and 1970. Limitations of space preclude a detailed consideration of all these documents here. Instead each will be examined briefly in terms of its most salient points, and then the implications of the 11 as a whole as an input to the debate will be considered. Of the 11 proposals, seven were from developed states, two from mixed or intermediate developed and developing states, and two from developing states; they will be examined below in these groupings.

(1) United States. The United States was the first in the field with its proposal 138/25 on August 3, 1970.[19] This document contained an elaboration of the position set down in the Nixon statement a few months before, and so reflected the attempt to find a suitable compromise among coastal, maritime, and international interests, both within and outside the United States. The proposal centered around the trusteeship zone idea as a compromise between wide and narrow coastal state limits. The zone was to stretch from a 200-meter/12-mile line out to a measured point of gradient on the continental rise. Within it the coastal state would act as the licensing and controlling agency, but would pay between one-half and two-thirds of all revenues received to the international seabed authority. The authority would consist of an assembly of all members, a council of 24 members, a secretariat subordinate to the council, a tribunal for settlement of disputes, and a number of commissions to oversee

specific areas like boundaries and operational practices. The council would be the dominant body with most powers of initiative, and would give a weighted vote to the developed states, the most advanced six of whom would always have seats. The assembly would have budget approval powers (that is, restraining capability) but few powers of initiative. Essentially the United States proposal was for a weak licensing type of regime and machinery, and the most distinctive aspect of it was the generous offer to share the very considerable revenues likely to accrue from oil under the U.S. continental shelf beyond 200 meters/12 miles in the relatively short-term future.

(2) Britain. The British proposal was also introduced on August 3, 1970 (138/26) and was elaborated in a second paper (138/46) on August 3, 1971. It was made in much less detail and in much shorter form than the American, but was similar to it in many respects. Britain supported the U.S. trusteeship zone proposal and advocated a weak licensing type of international regime and machinery. Its proposals for the structure of the seabed authority paralleled the assembly-council-secretariat-tribunal model offered by the United States, and suggested in addition an inspectorate to ensure the observance of technical standards. Britain opposed an authority with operational powers, and wanted to keep the international machinery simple by giving states considerable control over blocks of seabed to be allocated to them. Like the United States, Britain preferred to vest most powers of initiative within the authority in the council, and to make sure that the structure of the council gave particular weight to the interests of the developed states. Its proposal (138/46) noted the necessity "to make special provision on the Council for those States with an established sea bed technology," and contained an implicit threat in the subsequent observation that without the support of these states "no international regime in this field would be viable." As in the U.S. proposal, there was no dispute with the idea that seabed exploitation activities should be taxed in order to provide revenue from the common heritage, but there was concern that exploitation not be hampered and also that the machinery should be self-financing from seabed revenues. There was also concern that the convention itself contain the conditions of exploitation applicable to operators under the licensing system, although full elaboration of these conditions would have to await actual experience in commercial nodule mining operations.

(3) France. France made its proposal (138/27) on August 4, 1970, and like Britain's it was relatively brief. The French adopted an approach similar to the British one by basing their proposal on the idea of allocations of seabed blocks to states. Within these blocks states would have very wide discretionary powers over licensing and operations, and would even have considerable discretion

in the allocation of the benefits due to the common heritage. The French envisaged absolutely minimal international machinery, limited to a registry for mobile operations like nodule mining, and a state-dominated license system for fixed operations like oil drilling. The main function of the authority would be to allocate blocks of seabed to states, and the stress of the French proposal was to create maximum freedom for states, short of a national lakes solution, by keeping the role of the authority as limited as possible in regard to operations. The proposal said little about the structure of the authority except that it should have an assembly and a council, and that the latter should be the dominant organ. The idea of operational machinery was flatly rejected on grounds of lack of economic efficiency.

(4) Soviet Union. The Soviet proposal (138/43) of July 22, 1971 was brief and on most aspects extremely vague. It made no comment on the question of licenses, the question of distribution of benefits, or the problem of limits for the area. It did propose establishment of an assembly and a council, and awarded most powers of initiative to the latter, but was unclear as to what the relationship of the council to seabed exploitation should be, saying only that it should "supervise" such activities. The council itself was to decide questions of substance "by agreement", which seemed to imply a veto at least for the five regional groups into which the 30-member council was to be divided. The Soviet proposal was sufficiently ambiguous to make it difficult to classify with precision, but its general drift seemed to be toward a relatively weak international regime and machinery that would leave considerable initiative in the hands of states.

(5) Poland. Poland made its proposal (138/44) on August 3, 1971. Though brief, it was much more specific than that of the Soviet Union. The central idea was that the international machinery should start small and then grow in proportion to the need for it created by expansion of commercial activity on the seabed. The machinery would thus start with simple registry and coordinating functions and work its way up to some form of weak supervisory and licensing agency. The Polish proposal, like the Soviet, left most initiative on exploitation in the hands of states, and was careful to stress that "no group [of states] should be placed in a position in which it might be dominated by another group." To this end, it envisaged elaborate voting provisions with a general emphasis on decision-making by consensus. It contained mention of the "standard" structure for the authority consisting of assembly, council, secretariat, and possibly subsidiary bodies, but did not much elaborate on these except to suggest that the council would be the dominant organ. Poland also suggested fixing limits for the area, and offered

200 meters or 200 meters plus a fixed distance (not specified) as possible solutions.

(6) Canada. Canada presented its detailed, and in some respects quite innovative, proposal to the SBC on August 24, 1971. The Canadian paper (138/59) was based on a point-by-point consideration of the declaration of principles and reflected a pragmatic and compromise-oriented approach to the problem. The essence of its compromise was to combine the state initiative in exploitation favored by most developed states with some of the control elements (particularly with regard to regulating production, marketing, and distribution of seabed minerals) favored by many developing countries. The proposal stressed keeping international machinery to a minimum and encouraging private and state investment in exploitation in order to maximize the benefits from the common heritage. Thus it outlined a weak licensing regime (since most initiative would remain with states) but one with quite strong powers of supervision and control over standards, and with the possibility of an operational arm "at some future stage." The assembly and the council in the Canadian proposal appeared more balanced in power than in the council-dominated proposals of other developed states, and Canada expressed disapproval over the idea of weighted voting or group veto in the council. A secretariat, a tribunal, and various commissions were also provided for.

The Canadian proposal was notable for its emphasis on the rights and responsibilities of coastal states, suggesting in this regard both contributions to the seabed authority from seabed exploitation within the limits of national jurisdiction, and special rights for coastal states in a zone adjacent to national jurisdiction on the problem of pollution arising from seabed exploitation. Perhaps the most distinctive feature of the paper, however, was a detailed suggestion for transitional arrangements to cover the period before full-scale machinery could be established. The idea was to determine a "minimum non-contentious area" by having all states register their continental shelf claims (or the maximum beyond which they would not claim) and then set up a skeleton authority to license and supervise exploitation activities, to regulate production if necessary, and to collect revenues from the common heritage. This process was to be aided by an international development fund made up of voluntary contributions from states with exploitation income from their national continental shelf areas. The whole idea of transitional arrangements was seen as a means of preventing unilateral action by mining companies or states impatient with the international negotiations, and also as a way of preparing the ground for a full-fledged authority. It received little support because of fears both that it would encourage large continental shelf claims and that it would lead to a permanently truncated and "interim" international machinery.

(7) Japan. Japan's proposal (138/63) of November 23, 1971 was the last of the 11 to be submitted. In general it followed the trend set by earlier developed state schemes, except that it suggested a somewhat stronger seabed authority than most others. The structure of the authority was along the familiar assembly-council-secretariat-tribunal lines, with the council as the dominant body enshrining a privileged position for the developed states. The proposal was for a licensing type of authority to supervise the activities of states engaged in seabed exploitation, and much of it was devoted to suggesting details of licensing and conditions of exploitation. One of the few things distinguishing the Japanese proposal from those of most other developed states was the suggestion, made also by Canada, that "the Authority and the Contracting Parties shall pay due regard to the need for minimizing adverse effects of the development of the sea-bed resources on the prices of land-based minerals."

(8) Tanzania. The Tanzanian proposal (138/33) was the first to be made in 1971, and was introduced to the Seabed Committee on July 20. It was a relatively detailed document outlining an international regime and machinery of a much stronger type than any suggested by the developed states. Tanzania advocated a seabed authority with powers to operate as an exploiter itself, but also with powers to license states desiring to mine the international seabed. The authority was to distribute the benefits derived from the common heritage, and more important, it was to exercise complete control over the pricing and marketing of minerals produced from the international seabed. This latter function, designed to "balance the need of the world community for raw materials and the need for stability of the economies of the producers of land minerals, particularly when such producers are among the developing countries," was to be undertaken by a stabilization board within the authority. The structure of the authority was based on the assembly-council-secretariat model, but the assembly was to be the dominant body with most powers of initiative, and the functions of a tribunal were to be fulfilled by the ICJ. The Tanzanian proposal was authority-oriented rather than state-oriented, and appeared to give most powers of initiative in the area to the regime and machinery rather than to states. Since it offered a mixed system of state and authority exploitation, however, the proposal contained useful grounds for compromise.

(9) Chile, Colombia, Ecuador, El Salvador, Guatemala, Guyana, Jamaica, Mexico, Panama, Peru, Trinidad and Tobago, Uruguay, and Venezuela. The Latin American proposal (138/49) was placed before the SBC on August 10, 1971. Its 13 sponsors represented a good cross section of the Latin American regional group, including members from all regions (Central America, South

America, and Caribbean) and 200-mile as well as non-200-mile states. The only interest conspicuously absent was, as usual, that of the two landlocked states.

The proposal advocated a very strong seabed authority with wide-ranging and comprehensive powers. Building on a strict interpretation of the common heritage principle, the sponsors argued that all exploitation of the area should be undertaken by the authority. They rejected the idea of licensing as incompatible with the common heritage, and wished to restrict the role of states in exploitation to that of service contractor to, or joint venture partner with, the authority. In addition to exercising control over exploration and exploitation, the proposed authority would also undertake all activities relating to production, processing, and marketing; would provide for sharing of benefits from the common heritage; and would undertake, or supervise, all scientific research on the international seabed. It would strongly favor developing countries in all aspects of its function from the siting of facilities to taking "all necessary measures, including inter alia, control, reduction or suspension of production or fixing of prices of products obtained from exploitation of the area, whenever it deems that such production may have adverse economic effects for developing countries, exporters of raw materials." The authority would have an assembly-council-secretariat structure, with the assembly as the dominant body. It would also have a body to be known as the enterprise (this name can perhaps be seen as a riposte to the "trusteeship zone" of the United States), which would be empowered to "undertake all technical, industrial, or commercial activities relating to the exploration of the area and exploitation of its resources."

This proposal was the antithesis of most of those put forward by the developed states in that it gave virtually all powers of initiative to the regime and machinery instead of leaving them with states. As such, it defined the opposite end of the spectrum from the weak, state-oriented regime suggested by France, and so made clear the extent of the differences to be resolved in the ongoing negotiations in Subcommittee I.

(10) Malta. The Maltese proposal (138/53) was introduced on August 5, 1971. It was a lengthy and ambitious draft outlining not just a regime and machinery for the international seabed area, but for ocean space as a whole. It was largely the inspiration of Ambassador Pardo, and represented a second attempt by him to seize the initiative in ocean politics by injecting radical new ideas into the debate at a crucial point in time. In this instance he was attempting to take his original idea of an international seabed regime and machinery, which by 1971 had gathered considerable momentum, and marry it to the movement for a comprehensive conference on the

law of the sea, which had achieved success in Resolution 2750C in December 1970. His attempt was unsuccessful, among other reasons, because few people were willing to follow his conceptual leap, and because the regime he proposed compromised too many interests that were as yet too preoccupied with their own development to give much consideration to a grand compromise. Nonetheless, it represented a remarkably advanced and sophisticated attempt to design a good compromise on the law of the sea: an attempt that deserved more attention than it received and may yet come into its own.

There is not space in the present study to consider the whole draft ocean space treaty. Suffice it to say that it was based on three ideas: rejecting laissez-faire in the ocean space beyond the limits of national jurisdiction, rejecting the expansion of unfettered coastal state sovereignty, and expanding the concept of the common heritage to include all of international ocean space and not just the seabed. In terms of seabed resources, it gave coastal states exclusive rights to exploitation out to 200 miles or 200 meters (promising compensation to states losing margin areas thereby), but made this subject to revenue sharing with the international ocean space institutions. Beyond national jurisdiction, the proposal incorporated a mixed regime of licensing and exploitation by the institutions themselves, apparently giving emphasis to the former. The institutions would include an assembly, council, court and secretariat as well as several subsidiary functional organs, and Malta followed the developed states in giving most powers of initiative to the council. Avoidance of adverse effects on land-based mineral exporters was mentioned but not developed in detail, and no provisions were developed in detail for the machinery itself to undertake exploitation even although this seemed to be within its powers (Article 75). Malta, then, appeared to support a medium-strength licensing regime, although its proposals in this regard were insufficiently specific to make the nature of the regime wholly clear.

(11) Afghanistan, Austria, Belgium, Hungary, Nepal, Netherlands, and Singapore. The proposal of the seven landlocked and shelflocked states (138/55) was placed before the SBC on August 20, 1971, and presented a point of view rooted firmly in the needs and interests of such states. The group of sponsors itself was worthy of comment in that it contained no African, Arab, or Latin American states, and included a majority of developed states both Western and Soviet. This was in marked contrast to the list of sponsors of Resolution 2750B, which had been heavily dominated by African and Latin American states. A possible explanation of this contrast was that the African and Latin American landlocked states felt more constrained by regional group pressures on this issue than did their Asian and developed state counterparts. The proposals contained in

138/55 certainly went against the 200-mile trend evident in African and Latin American regional groups by this time.

The seven sponsoring states of 138/55 had no pretensions to present a full-scale draft for a comprehensive seabed regime, and were content to make points on aspects mostly of particular interest to landlocked and shelflocked states. They paid special attention to the question of the boundary for the area in an important move to create an alternative to the 200-mile position being pushed by the Latin Americans. The proposal was for a straight choice between 200 meters and 40 miles for the limits of coastal state jurisdiction over the shelf, with an additional 40-mile "priority zone" in which the coastal state would have veto rights over any exploitation under the auspices of the seabed authority. Under this scheme the maximum breadth for the territorial sea would be 12 miles. Such limits would leave significant oil deposits under the jurisdiction of the international seabed authority and would thereby make the common heritage a more substantial economic proposition—a policy advantageous to the landlocked and shelflocked since they had little to expect from extensions of coastal state jurisdictions.

The international seabed authority proposed by the seven states would be a mixed licensing and operating entity with the power to create its own operating agency. It would take steps to avoid, or compensate for, adverse effects from exploitation of the seabed on developing countries, and would facilitate access to the international seabed area for landlocked states involved in exploitation projects. Another novel proposal by the seven states was that in all organs of the authority where less than all members were represented (that is, the council), coastal and "non-coastal" (that is, landlocked or shelflocked) states should be represented in equal numbers.

These eleven proposals formed the backbone of SC.I's work for the period 1971-73. They contained all the elements of a seabed regime and machinery that achieved prominence, whether as points of agreement or as points of contention, and they defined the points of view and much of the membership of the major alignments on the subject. Although not all the proposals were strictly comparable, they fell into a rough spectrum ranging from weak regimes leaving most powers of initiative in the hands of states to strong regimes vesting most powers of initiative in the international machinery. The proposals of France, Poland, and the Soviet Union occupied the weak end of the spectrum; those of Britain, Japan, and the United States were clearly still on the weak side, but somewhat more toward the middle; those of Canada and Malta (disregarding the ocean aspects of the Maltese proposal) were compromise proposals aimed at the center ground; those of the seven landlocked and shelflocked states and of Tanzania were clearly on the strong side but still contained elements of compromise; and that of the 13 Latin American states occupied the extreme strong end of the spectrum. After all the proposals were in, it was evident that there was no significant change in the fundamental developed versus developing alignment that had dominated this issue from the early years of the SBC.

On a general level the proposals and the subsequent debates revealed a certain measure of agreement over and above that achieved in the declaration of principles. Most states, for example, accepted the basic assembly-council-secretariat-tribunal-commissions structure for the seabed authority, even though there were substantial differences of opinion on the powers and functions of these structures. In general, however, it was the areas of disagreement that dominated the discussion.

Although the discussion in SC.I ranged over a large number of topics and details, certain issues tended to emerge as focal points in the confrontation between developed and developing countries. These issues represented the crux of the differences underlying the alternative conceptions of the purposes and powers of the international regime and machinery, and as the debate progressed, they became more and more prominent. It became increasingly clear through 1972 and 1973 that all the peripheral matters awaited settlement of these central differences, and consequently that if these relatively few points could be settled, other matters should fall into place quite quickly. The effort to identify, define precisely, and find compromises for these key points became the central theme of SC.I's work.

Other than the question of limits, which did not follow a developed/developing alignment pattern, three issues can be identified as crucial: the nature of the operating system for seabed mining in the international area, the location of powers of initiative within the international machinery, and the question of whether or not the seabed authority should have powers to regulate production in order to prevent disruptive competition with land-based metal producers.

The nature of the operating system was the first key issue to emerge, and, as illustrated in the previous chapter, this had already stimulated visible alignments by the end of 1970. (See Table 6.3.) The rush of proposals in 1970-71 greatly clarified the alternative possibilities by outlining the four types of operating systems: a registry (France, Poland); a weak licensing system (Britain, United States, Japan, Poland?, Soviet Union?); a mixed licensing/authority operating system (Malta, Canada, the seven landlocked and shelf-locked states, Tanzania); and a pure authority operating system (the 13 Latin American states). Most of the proponents at either extreme of this spectrum rejected the position of their opponents at

TABLE 7.1

Expressed Opinion on a Seabed Authority With Direct Powers Over Exploitation, 1969-73

Opposed[a]	In Favor With Reservations[a]		In Favor[b]	
France	Australia	Spain	Brazil	Algeria
United Kingdom	Canada		Venezuela	Botswana[c]
United States	Denmark[d]	Austria[c]	Peru	Burundi[c]
Japan	New Zealand	Belgium[d]	Chile	Cameroon
Italy		Netherlands[d]	Uruguay	Central African Republic[c]
Soviet Union		Hungary[c]	Trinidad and Tobago	Chad
Czechoslovakia[c]			Panama	Congo[c]
Bulgaria		Yugoslavia	Ecuador	Dahomey
Poland[d]			Colombia	Equatorial Guinea
			Mexico	
Lebanon			Jamaica	Ethiopia
			El Salvador	Gabon
Philippines			Guatemala	Gambia
			Barbados	Ghana
			Guyana	Guinea
			Costa Rica	Ivory Coast
			Nicaragua	Kenya
			Haiti	Lesotho[c]
			Dominican Republic	Liberia
				Libya
			Honduras	Malagasy Republic
				Malawi[c]
			Kuwait[d]	Mali[c]
			Iraq[d]	Mauritania
				Mauritius
			Sri Lanka	Morocco
			India	Niger[c]
			Afghanistan[c]	Nigeria
			Nepal[c]	Rwanda[c]
			China	Senegal
			Singapore[d]	Sierra Leone
			Turkey	Somalia
				Sudan
				Swaziland[c]
				Tanzania
				Togo[d]
				Tunisia
				Uganda[c]
				Egypt
				Upper Volta[c]
				Zaire[d]
				Zambia[c]

[a]Statements made in the SBC

[b]Adherence to the OAU Declaration of May 1973 or the Declaration of Santo Domingo; or sponsorship of A/AC.138/33, 49, or 55; or statements made in the SBC.

[c]Landlocked

[d]Shelflocked

Note: If a state made conflicting statements on this issue, the latest statement was used to determine its position above.

the other end, while those in the middle tried to establish grounds for compromise. The dominant question here was a licensing versus an operating authority, and the respective underlying viewpoints were desire for secure access to seabed minerals versus fear of looting of the common heritage to the benefit of the developed states. Most of the developed states feared that a large international machinery with wide powers would not only be hopelessly inefficient in economic terms, but also would discriminate against their interest in obtaining a new source of minerals, and would itself absorb most of the revenues generated from the common heritage. The developing countries hoped that such machinery would not only prevent the developed states from taking advantage of their technological powers, but also would give the poorer states a chance to participate in a major economic venture. To an increasing extent also, they looked upon the seabed authority as a new model in international economics that would be designed to rectify the imbalance between themselves and the developed states. The question of licensing versus operating machinery thus became a rallying point around which the complex differences between developed and developing countries could be expressed in simple, almost symbolic, terms.

The symbolic aspect of this issue dominated the discussions in SC.I throughout 1972 and 1973, so that the debate was more a relatively static process of building up support on one side or the other than a dynamic process of negotiating towards a compromise. By 1973, however, the process of delineating sides had gone about as far as it could go, and indeed had become extremely repetitive and unproductive. Since the great majority of states (the developing countries) eventually favored an operation authority along the lines proposed by the Latin American states, the debate degenerated into a collection of statements of those in favor of, or opposed to, such an authority.

Only toward the end of the SBC's life did a new line of negotiation open up to offer prospects for some kind of movement. This was introduced by developed states, and had been in the background in many of their proposals from 1970-71. It concerned the actual legal conditions of exploitation, regardless of whether they were called a license, a joint venture, or a service contract, and as such tended to move the discussion to a more specific level. Instead of talking fruitlessly about licensing versus operating authorities, these states wanted to talk about the specific rights and duties of operators and the specific rights and duties of the authority in regard to such matters as fees, financial risks, areas to be allocated, work requirements, integrity of investments, operating standards, and duration of contracts. This shift was becoming visible in the last years of the SBC, but did not become dominant. It was made partly in response to the growing paralysis of the debate on the more general level, and partly in response to the grudging recognition by the developing countries that any nodule mining activity would have to utilize the technology of the developed states at least for a substantial number of years. "Conditions of exploitation," however, did not fully replace "licensing versus operating" as the core of the debate on the nature of the operating system until the Caracas session of the Third Law of the Sea Conference in the summer of 1974.

The second key issue, the location of powers of initiative within the international machinery, was essentially a debate over whether the assembly or the council should be the dominant organ of the seabed authority. This arose out of the attempt of the developed states to protect themselves as a minority by giving power to the council in their proposals, and out of the parallel attempt of the developing countries to protect themselves from the technological advantage of the developed states by giving power to the assembly, which they could control through their votes. The issue was divided along developed/developing lines except for Canada, which took a more or less even-handed approach.

The question of assembly versus council, like that of operating versus licensing, also had considerable significance in the developed/developing context in terms of "equal rights for all sovereign states" versus "privileged status for the wealthy minority." To a large extent, however, emphasis on the question reflected a particular stage in the debate. In 1971, and to a lesser extent in 1972, SC.I. tended toward establishing the seabed authority with only a very general mandate as to what it should actually do. The idea was that the authority should be able to work out its role on the basis of experience, and should therefore be relatively flexible in its early years. Such an approach obviously placed a great premium on where power resided within the structure of the organization, because whichever organ controlled the powers of initiative would be able to determine in what ways the authority would develop. After 1971, however, this approach receded in favor of giving the authority a fairly specific mandate (licensing versus operating, conditions of exploitation), and so discussion of the actual structures of the seabed authority dwindled into a peripheral concern.

The third key issue, regulation of seabed mineral production by the authority, was largely but not wholly related to the question of a licensing versus an operating authority. The basic conflict here was between the fears of the developed states that such controls would render nodule mining economically unattractive and the fears of the developing countries that without such controls exploitation of the common heritage could turn into a curse rather than a blessing by undercutting the metal exports of some of their number.

Considering the relatively few among their ranks that were heavily dependent on exporting copper, nickel, cobalt, and manganese, the solidarity of the developing countries was quite remarkable, and showed again the strong symbolic power of these issues in the developed/developing context. As the prospect of significant financial gain from the common heritage declined, the developing countries put more and more emphasis on controlling possible negative effects from seabed exploitation. Aside from the extremely difficult problem of how such effects could be identified and alleviated, the strong demand for production controls put the developing countries into a position that at times looked like flat opposition to any exploitation of the common heritage. This tended to create an impasse in the negotiations with those developed states anxious to undertake seabed mining, and was not a tenable position given the option of the developed states to mine unilaterally. It also tended to put the developing countries out of phase with the original impetus of the SBC to operationalize the common heritage by creating a new and innovative type of international economic organization to exploit a common resource.

Very little progress towards a solution of this problem was made in the SBC, partly because of the inherent difficulty of devising solutions, partly because of profound disagreements over the seriousness of the problem, and partly because of the strong position of both sides in the negotiations. As already discussed above, disagreement over the likely impact of nodule mining ran deep, and since no hard data on the subject existed, the speculative attempts at forecasting had become politicized and absorbed into pre-existing opinions. The developed states argued that the impact would be small and the developing countries that it would be large, and so both kept in line with their general positions respectively in favor of weak and powerful international machinery.

In addition to these three key issues at the center of the debate in SC.I., two other issues achieved noteworthy prominence. The first of these was the proposal made by Canada, the United States, and Malta in their respective seabed regime proposals that revenue from the continental shelf within national jurisdiction should be shared with the seabed authority. This proposal related closely to the question of limits, and will be discussed under that heading in the next chapter.

The second issue concerned the right to conduct scientific research on the international seabed, and centered on the question of whether research in that area should be a free right as it had always been or should be subject to regulation or control by the seabed authority. Those in favor of research remaining free included the United States, the Soviet Union, Britain, and Malta, and they argued that states were the only source of technological and economic capabilities to undertake this necessary but not directly remunerative task. Canada favored regulation of research by the authority mostly on the grounds of environmental hazards arising from such research practices as deep drilling. The 13 Latin American states, in line with their comprehensive approach to the authority, wanted it to have the power of authorization as well as that of regulation.

This issue was in many respects merely an annex to the much more hotly contested question of scientific research within the coastal zone. The prevailing alignment was developed versus developing, and the issue was only one of the many components in the more general argument over whether the seabed authority should be large and comprehensive or small and limited. No reconciliation of the two sides was in sight by the end of the SBC's work in 1973.

What, then, was the end result of SC.I's work?

SC.I started with a considerable advantage over the other two subcommittees in that the old SBC had already been working on the international seabed regime question for three years. Despite this head start, however, SC.I failed by a large margin to agree upon a unified set of draft articles. It spent 1971 completing a wide-ranging debate based on the 11 regime proposals considered above. In 1972, aided by a comparative table of the proposals drawn up by the Secretary-General (138/L.10), the subcommittee moved to more specific discussion of particular items and set up a working group to try to reduce the numerous alternative texts to single agreed draft articles.[20] In 1973, the subcommittee continued its discussion of particular items, and the working group continued its efforts to reduce the number of alternative texts. Despite nearly 100 meetings, however, the working group did not produce a single agreed draft article, and many of the items on its list were still represented by three or more alternative texts. The final result of its work was 115 pages of alternative texts covering nearly 100 items,[21] a result in stark contrast to the unified draft convention presented to the First Law of the Sea Conference by the International Law Commission.

The list of alternative texts was appropriately titled "texts illustrating areas of agreement and disagreement," and this about summed up the significance of the SBC's six years of work on this issue. The 90 members of the committee had reached no conclusions at all by the end of 1973, but were still in the middle of their task when their mandate to continue expired. They had made good, but not complete, progress in defining the subject (the question of conditions of exploitation was not covered in the list of texts, and therefore some states refused to accept the list as complete), and, as discussed above, they had identified and started refining the key points of disagreement. But the state of their work offered no grounds for optimism about the conference for which their report was supposed to be the preparatory document. A careful examination of the alternative texts produced by SC.I reveals that in fact there was no meaningful advance on the level of agreement achieved in the declaration of principles. In terms of advancing toward consensus, SC.I had been paralyzed for the entire three-year period of 1971-73. What it had achieved was a substantial definition, clarification, and elaboration of areas of difference, an achievement representing only the first step toward the level of agreement necessary for a draft convention.

Seven main reasons stand out as responsible for SC.I's lack of progress. They were all operative from the beginning of the SBC, and have been constant themes in Chapters 4 through 7. First was the lack of a definition of the area to which the regime was to apply. SC.II had reached no agreement on this by the end of 1973, and therefore the nature of the resources appertaining to the international regime remained unclear. Second was the innate slowness of the consensus procedure when used among a large number of states. This was exacerbated by the great influx of new states into the SBC in 1971. Third was the genuine complexity of the issue, involving the necessity to design a new type of international organization to govern an unprecedented activity with few known characteristics Factors such as the politicization of information and the spillover into other issues like commodity policy and coastal state jurisdiction added to this complexity. Fourth was the technological and economic development of seabed mining, which brought the prospect of commercial exploitation ever closer and raised political tensions around the issue while lowering general economic interest in the possibility of wealth for the developing countries. Fifth was initiative-taking by, or within, states in response both to economic and technological developments and to developments in the international negotiations. Examples of this were the consideration of a unilaterally declared legal regime for nodule mining in the United States and the diplomatic offensive by the Latin American states on behalf of a 200-mile territorial sea.[22] Its effect was to complicate SC.I's work by emphasizing points of conflict. Sixth was the highly polarized and evenly balanced alignment in SC.I of developed versus developing countries. The rigidity of this alignment, with all its symbolic associations external to the seabed issue, discouraged compromise movements by giving each side a strong base to fall back on. The process described above reflected such a consolidation of positions, and the efforts of the very few states not bound up in the polarization (Canada, Malta) were unable to prevail against it. Seventh and finally was the shift of interest to coastal zone matters, which decreased the importance of and attention devoted to the international seabed area.

Given that the effect of these factors was to slow the work of SC.I to the point of paralysis, what was the impact of the lack of progress on the state of the issue itself, and consequently on the international initiative it represented? Unquestionably the most important impact was the decline in the importance of the seabed regime issue both absolutely and in relation to other issues. Enthusiasm for and optimism about the seabed regime waned as its economics became less certain and as the problems associated with implementing it grew more formidable. Concern shifted to the more immediate and concrete issue of coastal state control over the continental shelf and coastal zone—an issue originally stimulated by the boundary question of the international regime. The continuing delay in implementing an international regime likewise strengthened the hand of those states engaged in developing unilateral options on deep ocean mining. The result was that by the end of 1973 the international initiative was still a long way from consummation, and was in danger of being shelved or abandoned in favor of other issues (where coastal states had the initiative) or of being overtaken by the initiatives of developed states interested in nodule mining.

In any event, the problem no longer lay with the Seabed Committee but had been passed on to the more highly pressurized forum of an international conference.

NOTES

1. GAOR XXVI, Suppl. 21, A/8421, September 1971 (Report of SBC), p. 5.

2. The following section is heavily indebted to Arnold J. Rothstein and Raymond Kaufman, "The Approaching Maturity of Deep Ocean Mining—The Pace Quickens," Offshore Technology Conference 1758; and a letter of June 7, 1973 from T. S. Ary to Senator Lee Metcalf. Both are reprinted in U.S. Senate, Mineral Resources of the Deep Seabed, 93rd Congress, first session on S.1134. See also UN Doc. A/AC.138/90 (July 1973).

3. Business Week, June 16, 1973.

4. Mineral Resources of the Deep Seabed, op. cit., pp. 2-14.

5. Ibid., pp. 121, 232-33.

6. Ibid., pp. 110-11.

7. Ibid., p. 112.

8. U.S. Senate, The Law of the Sea Crisis (Washington: Government Printing Office, 1972), p. 10.

9. Mineral Resources of the Deep Seabed, op. cit., pp. 36-40, 230.

10. Letter of March 1, 1973 from Charles N. Brower to Senator H. M. Jackson, reprinted in ibid., pp. 15-23.

11. Ibid., pp. 15-23, 242-43.

12. A/AC.138/SC.I/SR.64, pp. 21-22.

13. The sponsors were Algeria, Brazil, Chile, China, Iraq, Kenya, Kuwait, Libya, Mexico, Peru, Venezuela, Yemen, and Yugoslavia: GAOR XXVII, Suppl. 21, A/8721, 1972 (Report of the SBC), p. 69. Supporters included India, Sri Lanka, Pakistan, Cameroon, Nigeria, and Mauritania: A/AC.138/SR.76, p. 60.

14. Report of the SBC, ibid., p. 69.

15. Ibid., pp. 9-11; A/AC.138/SR.79, pp. 20-21 and SR.83, p. 63; and A/AC.138/SC.I/SR.49-50.

16. Letter from Bower to Jackson, op. cit.

17. The argument ran that if the total world demand for any one of the four nodule metals was taken as a production target, the other three would be produced in amounts disproportionate to total demand. For example, production of 100 percent of world nickel demand from nodules would involve production of 138.2 percent of world demand for manganese and 543.7 percent of world demand for cobalt. A somewhat different set of figures on this problem was produced in F. L. Laque, "Deep Ocean Mining: Prospects and Anticipated Short-Term Benefits," in Borgese, ed., Pacem in Maribus (New York: Dodd, Mead, 1972), p. 144.

18. For other considerations of possible benefits, see Laque, ibid., and Raul Branco, "The Tax Revenue Potential of Manganese Nodules," Ocean Development and International Law Journal 1, no. 2 (Summer 1973): 201-08.

19. Reprinted in GAOR XXV, Suppl. 21, A/8021, 1970 (Report of the SBC), pp. 130-76. All of these proposals except that of Japan were reprinted either in this report or in the one for the following year: GAOR XXVI, Suppl. 21, A/8421, 1971. Japan's is reprinted in Oda, The International Law of the Ocean Development (Leiden: Sijthoff, 1972).

20. The list of alternative texts was compiled from all the proposals made by states.

21. GAOR XXVIII, Suppl. 21, A/9021, 1973 (Report of the SBC), Vol. II, pp. 141-66.

22. An interesting view of the strong Latin American seabed regime proposal was to interpret it as a mere adjunct to their coastal state campaign. The logic would be that by creating Group of 77 support for a strong regime, the Latin American coastal states would be able to make it sufficiently distasteful to the developed states, particularly the United States, to force them to adopt wide coastal state jurisdiction in self-defense. This would undermine maritime state opposition to the Latin American claims and consoli-

date the 200-mile territorial sea position. Such an interpretation assumes a low priority attitude on the part of the Latin Americans toward the seabed regime (which is questionable given Chilean and Peruvian copper interests) and a depth of cynicism and deviousness well beyond that normally associated with international negotiations. This interpretation was, nevertheless, made in some quarters. (Interviews with delegates.)

CHAPTER

8

SEABED POLITICS AND THE LAW OF THE SEA, 1971-73, PART II: DEFINING THE BOUNDARY BETWEEN NATIONAL AND INTERNATIONAL SEABED AREAS

The effort to define the boundary between national and international seabed areas took place against a background of pressure from three sources. First, ongoing developments in the offshore oil industry continued to raise the value of the continental margin and thereby to increase the interest of coastal states in establishing control over seabed areas adjacent to their coasts. Second, there was considerable disparity among the various national claims on the question of limits and no agreed interpretation of international law on the matter. And third, there was a very substantial movement, led initially by the Latin American coastal states, to link the seabed limits issue to the question of coastal state jurisdiction over offshore maritime areas as a whole, including both continental margin and water column. This movement complicated the seabed limits question enormously by tying it to the traditionally controversial mass of issues involved in coastal state jurisdiction over the high seas.

SEABED RESOURCES: DEVELOPMENTS AND DISPUTES

The headlong expansion of the offshore oil industry continued unabated throughout the period 1971-73. The industry sustained its advance into new offshore areas so that by 1973 the only major shelves not penetrated by Western oil companies were those off the Soviet Union, China, and Antarctica. New technologies facilitating the push into ever deeper waters included dynamic positioning systems for drilling vessels, systems to enable hole re-entry at great depths, and wellhead production systems capable of operating on the seabed rather than on a surface platform.[1] Production depths increased steadily, with many wells between 400 and 500 feet under preparation in the North Sea and expectations of production from depths up to 1,000 feet by the end of the decade. In effect, there were no longer any significant technological constraints on continental shelf oil exploitation, but only constraints imposed by the higher costs of deeper drilling.

By 1973, commercial production of offshore oil was under way off the shores of 26 states. Commercial discoveries had been made off the shores of a further 13 states, but were not yet in production, and exploration was under way off the shores of a total of 80 states.[2] Eighteen percent of world oil production and 10 percent of world gas production came from offshore sources in 1972, and the total annual volume of offshore oil produced had risen 27 percent between 1970 and 1972 from 2.6 to 3.3 billion barrels.[3] The offshore oil boom continued to fuel not only the industries of the developed states but also the aspirations of coastal states both developed and developing.

The boom, as discussed in Chapter 6, opened up a wealth of possibilities for conflict between states. None of these disputes broke into armed conflict during the life of the SBC, but a new area of dispute arose in Antarctica. American scientists working under the auspices of the National Science Foundation conducted geological surveys of the continental shelf around Antarctica during 1973 and discovered indications of substantial oil deposits. Their results created pressure for an exploitation program in the area which if carried out would have a disruptive impact on the 17-nation treaty restricting national ownership of any part of Antarctica.[4]

The steady trickle of continental shelf delimitation agreements also continued under pressure from the offshore oil boom, but remained confined to a few relatively closed areas. In 1971 West Germany, Denmark, the Netherlands, and Britain finally settled their shelf boundaries on the basis of the 1969 ICJ decision, thus completing the demarcation of the shelf underlying the North Sea. In the Mediterranean, Italy and Tunisia agreed on a shelf boundary, while in the Far East, Australia and Indonesia, and Indonesia, Malaysia, and Thailand added further to earlier settlements on the shelf surrounding Indonesia. Canada and Denmark reached agreement on the Greenland-Canada delimitation in December 1973.

The question of seabed mineral resources was, of course, central to the dispute over limits between national and international jurisdiction. In 1972 it provided the basis for a major confrontation between coastal states favoring wide national limits and landlocked and shelflocked states favoring narrow national limits. The immediate cause of dispute was a document (138/81) submitted to the SBC by a well-balanced group of 11 landlocked and shelflocked states (three Asian, three Western, two Eastern European, one Latin American, and two African) requesting a study "on the different economic implications of the various proposals for limits of the International Seabed Area." The 11 sponsors suggested that the study consider the implications of limits set at 200 meters, 500 meters, the edge of the margin, 40 miles, and 200 miles; the purpose of their request was to obtain precise information on the quantity and type of resources that were hanging in the balance between national and international control. Since all of the sponsors of the seabed regime proposal 138/55 (see Chapter 7) that had favored narrow limits were among the 11, the request was clearly both a response to the trend toward wide claims and a statement in favor of more resources for the international area.

Predictably, the Latin American coastal states led the opposition to the request, arguing, among other things, that such a study was beyond the resources of the Secretary-General, that it would prejudice the position of states with wide claims, and that the five limits specified were not acceptable.[5] Afghanistan, in reply, noted the problems of the landlocked developing countries in maintaining solidarity with other developing countries if they were not given equal treatment.[6]

The dispute was carried over into the 1972 meetings of the General Assembly, where it centered on Resolutions 3029B and C. Resolution 3029B, sponsored by 31 landlocked and shelflocked countries (five Western, three Soviet group, two Latin American, 13 African, three Arab, and five Asian), requested "a comparative study of the extent and the economic significance, in terms of resources, of the international area that would result from each of the various proposals on limits of national jurisdiction submitted [to the SBC]." Such a study would inevitably emphasize the adverse effect of wide national claims on the resources available to the common heritage. Counterposed to it was Resolution 3029C, sponsored by Peru, requesting a parallel study on the economic significance of different limits for coastal states.

The voting on these two requests provided further evidence of the essentially landlocked and shelflocked state versus coastal state nature of this dispute. Resolution 3029B passed by 69 to 15 to 41. Twelve of the 15 opposing votes were cast by Latin American coastal states, and all 15 were cast by states that had expressed support for a 200-mile limit. Only two of the 41 abstentions were by landlocked or shelflocked states, while half were by states favoring 200 miles, and many were by prominent coastal states such as Australia, Canada, Norway, New Zealand, India, Indonesia, the Philippines, Sri Lanka, Mexico, and Senegal. Resolution 3029C passed by 100 to 0 to 28, and 12 of the abstentions were by landlocked and shelflocked states.

The product of these two resolutions was the single report of the Secretary-General (138/87) that was discussed in Chapter 7. The report demonstrated clearly that the difference between wide and narrow limits involved the question of whether or not the common heritage would include commercially significant oil resources. The significance of the battle over the study was that it marked a counter-offensive by the landlocked and shelflocked states against the Latin American-led coastal state movement. In effect, the 31 sponsors of Resolution 3029B were making a declaration of both their demand for equal treatment in any redistribution of ocean resources and their voting strength in the forthcoming conference. They were exerting pressure for consideration of their interests not only on the international community as a whole but also on the regional groups and the Group of 77. One result of their vigorous campaign on the limits/resources study was to give the impression that wide versus narrow limits would be a central conflict at the Conference. In fact that did not turn out to be the case. Well before the end of the SBC's life, movement was under way in the regional groups and the Group of 77 to try to accommodate the demands of the landlocked and shelflocked states within a 200-mile framework, and thereby bring them back into the fold.

REGIONAL ACTIVITY AND NATIONAL CLAIMS

The period 1971-73 saw a continuation and expansion of the regional activity external to the SBC on the law of the sea begun in 1970 by the two Latin American declarations and the statement of the nonaligned countries. Most of this activity, following the Latin American lead, concentrated on the issue of the limits and nature of coastal state jurisdiction. Nearly all of it took place among the developing countries, and despite the numerous separate bodies in which it occurred, there was an unmistakeable trend toward some form of 200-mile limit.

One of the first results of Resolution 2750C (1970, setting a date for a new law of the sea conference) was the adoption by the Asian-African Legal Consultative Committee (AALCC) of the law of the sea as a priority item for discussion at its 12th session in January 1971.[7] The AALCC was not a regional group in the same sense as the African and Latin American groups, since its primary function was to discuss legal problems rather than to formulate regional policy positions. Nevertheless, its discussions played an important role in that they helped to inform and stimulate the policies of a key group of developing countries. Both the United States and the Latin American countries sent observers to AALCC sessions in order to present their own favored policy line to the assembled delegates.

The 1971 session took up the law of the sea at the urging of Indonesia, and the discussion ranged over nearly all the topics included on the agenda of the new Law of the Sea Conference. No conclusions were reached, but substantial support was found for a 12-mile territorial sea with an adjacent exclusive fishing zone of unspecified dimensions, and for an international seabed regime and machinery along mixed operating/licensing lines with powers to prevent adverse effects on land-based producers. The committee found less evidence of agreement on the boundary between national and international jurisdiction on the seabed, although more states supported 200 miles than supported 200 meters. Indonesia and the Philippines conducted an energetic campaign in favor of archipelagic states having the right to claim the seas enclosed by their island group; Ecuador, Brazil, Argentina, and Peru presented their case for a 200-mile territorial sea; and the United States argued for the compromise position laid down in the Nixon statement and in 138/25. None of this lobbying resulted in much support. The United States proposal in particular came under fire for the fact that the "trustee" appeared to gain more than the "beneficiary". Somewhat surprisingly in the light of subsequent moves toward a 200-mile seabed boundary, the American proposal was also criticized for being incompatible with the common heritage.[8] In retrospect, probably the most important aspect of the AALCC's 1971 session was the development of thinking along economic zone lines that arose naturally out of the discussion of seabed and fisheries limits for national jurisdiction.[9]

Of greater significance than the AALCC discussions was a set of three resolutions on law of the sea matters passed by the Council of Ministers of the OAU at its session in June 1971. These resolutions marked the full-fledged emergence of the African group as an actor in law of the sea negotiations, an emergence that had been strongly foreshadowed by the Lusaka Statement in 1970.

The first resolution drew attention to the issue of seabed exploitation in general, and requested a study by various OAU subsidiary organs with a view to harmonizing African positions and presenting a united front at international meetings. The second resolution argued for linking coastal jurisdiction over the continental shelf to coastal jurisdiction over fish resources in the superjacent waters, a policy reminiscent of Latin American efforts in the 1940s and 1950s. It urged the African states to extend their sovereignty over the resources of the waters superjacent to the shelf without proposing a definition to cover limits, and it urged regional cooperation in the development of fisheries without specifying whether or not this was to include landlocked states as well as coastal states. The third resolution bore only indirectly on seabed matters. It reaffirmed the right of states to exercise permanent sovereignty over their natural resources, and, more significantly, it called for the formation of an African union of mineral-exporting countries. This latter move provided further illustration of the efforts of regional groups and the Group of 77 to co-opt the mineral-exporting states by adopting their demand for protection against competition from exploitation of the international seabed area.[10]

In addition to their significance as a milestone in the development of an African regional group on the law of the sea, the three resolutions were also significant for their content. Although they were all generally worded and lacking in detail, they demonstrated an overwhelming concern with states' rights that contrasted markedly with the international regime orientation of the Lusaka Statement. Between 1970 and 1971, the African states had clearly shifted their concern from the international area to the coastal zone. The broadening of the SBC's mandate to include the whole law of the sea can be seen as partly responsible for this shift, but declining optimism over the prospects of an international regime, and influence from the Latin American coastal state offensive, must also be counted as important factors.

Further evidence of the swing away from interest in the international seabed regime and toward preoccupation with coastal state rights came with the resolution on marine resources adopted by the ministerial meeting of the Group of 77 in November 1971—held, appropriately, in Lima. The first part of the resolution, though shortened and weakened, bore an unmistakeable resemblance to the unilateralist sentiments in the Montevideo and Lima Declarations of 1970:

> The Second Ministerial Meeting of the Group of 77 . . . decides: to affirm as a common aim of the Group of 77 recognition by the international community that coastal States have the right to protect and exploit the resources of the sea adjacent to their coasts and of the soil and subsoil thereof, within their limits of national jurisdiction, the establishment of which must take due account of the development and welfare needs of their peoples.[11]

Justification for this position was found in the argument that "some developed states" were using their scientific, technological, and financial advantage to exploit marine resources off the coasts of other states. This part of the resolution reflected the success of the developing coastal states, with the Latin Americans in the vanguard, in establishing their influence within the Group of 77. It also illustrated the success with which these states had strengthened their policy by associating it with the general anti-developed state position of the Group of 77. By linking the overall coastal state position to the image of developed states plundering the resources off the coasts of developing countries, the developing coastal states were able to achieve a substantial measure of support for their position within the Group of 77 despite the fact that coastal state policy held no direct advantage for many developing countries.

The same tendency of the Group of 77 to co-opt active subgroups by adopting their position was illustrated in another part of the resolution. This supported minimization of "any adverse economic effect caused by the fluctuation of prices of raw materials" resulting from exploitation of the international seabed area, and so took into account the interest of the mineral-exporting states—a minority group also successful in linking its policy to the anti-developed state line and subsequently successful in gaining Group of 77 support despite conflicting interests with other developing countries.

This Group of 77 resolution was the only important activity on the law of the sea undertaken by the Group outside the context of the SBC. It indicated the Group's concern to build a common front of developing countries on law of the sea issues and its ability to absorb the positions of particular subgroups in order to maintain unity. More important, however, was the evidence in the resolution that the Group of 77's attention as a whole was shifting toward the coastal zone issue and away from the international seabed regime.

The AALCC resumed its discussion on the law of the sea at its 13th session in January 1972. The meeting attracted observers from 38 states from both within and outside the Afro-Asian area, including delegations representing the United States, the Soviet Union, and Britain. As in the previous year, the AALCC devoted itself to discussion rather than policy formation, and so did not produce any conclusions. The agenda again included nearly all issues under negotiation in the SBC. A number of working papers were introduced, including one from Sri Lanka on the international seabed regime,[12] one from Kenya on the exclusive economic zone, one from Japan on high seas fisheries, one from Indonesia on archipelagic states, and one from Malaysia on international straits. Despite the major paper from Sri Lanka on the international seabed regime, the emphasis of the session was on the coastal economic zone, including fisheries, and a "great majority of developing countries" favored the economic zone concept.[13] The debate on the boundary between national and international jurisdiction on the seabed was inconclusive, and the prominence of Kenya's paper as opposed to that of Sri Lanka underlined the strength of the ongoing shift toward coastal concerns.

In June 1972, another important regional declaration came out of a meeting of the Central American and Caribbean states. Known as the Declaration of Santo Domingo, it was the end product of a series of meetings among these states dating back to November 1971. The declaration covered most major law of the sea topics, but its centerpiece was a new formulation for coastal state jurisdiction known as the "patrimonial sea". The patrimonial sea was to extend from the outer limit of a 12-mile territorial sea to a limit not exceeding 200 miles from the points from which the territorial sea was measured. Within this area, the coastal state would have sovereign rights over all resources including seabed resources, and would also have the right to "regulate" scientific research and to adopt "the necessary measures" to prevent marine pollution. The difference between a patrimonial sea and a territorial sea would be that in the former the traditional high seas rights of freedom of navigation and freedom to lay cables and pipelines, would be maintained, though subject to "restrictions . . . resulting from the exercise by the Coastal State of its rights within the area."[14] For the continental shelf, the Santo Domingo states proposed maintenance of the 1958 convention definition, or possibly its replacement by a fixed limit "taking into account the outer limits of the continental rise." Thus, in cases where the continental margin extended beyond 200 miles, the coastal state would retain the rights to seabed resources provided for in international law—that is, sovereign rights to explore and exploit.

For the area beyond the limits of national jurisdiction, the Santo Domingo states proposed a strong operational machinery "empowered to undertake all activities in the area." Although this aspect was not developed in much detail in the declaration, it appeared to be generally in line with the Latin American seabed regime proposal (138/49) put forward in the SBC the previous year. In addition, they proposed that high seas fishing should not be "unrestricted", but should be "the subject of adequate international regulation." This proposal also was not developed in much detail, and was not explicitly linked to the powers of the international seabed authority. Regional cooperation was proposed for the Caribbean area, but no mention was made of the more general problems of landlocked and shelflocked states.

The patrimonial sea proposal bore an obvious relation not only to earlier Latin American coastal state claims (the epicontinental sea of Argentina and the maritime zones of Chile, Ecuador, and Peru), but also to the whole economic zone idea arising from the discussion in the SBC. Of the 15 states attending the Santo Domingo Conference, ten supported the declaration (Colombia, Costa Rica, Dominican Republic, Guatemala, Haiti, Honduras, Mexico, Nicaragua, Trinidad and Tobago, and Venezuela) while five abstained (Barbados, El Salvador, Guyana, Jamaica, and Panama).[15] Ecuador, among others, was careful to dissociate itself from the declaration, making clear that it did not reflect the position of all Latin American states, but only of the ten states that voted for it.[16]

Venezuela, which along with Colombia and Mexico had been one of the prime movers behind the patrimonial sea idea, argued that the Declaration of Santo Domingo was an attempt to achieve a common Latin American position, and not to set up a rival group to those established earlier.[17] Inasmuch as all the signatories of the three Latin American declarations agreed on a 200-mile limit for national jurisdiction in the water column and on the edge of the margin as the limit for national jurisdiction on the seabed, the majority of Latin American states did advocate a common position. But within this agreed limit, important differences remained. Only Nicaragua adhered to all three declarations, and in effect the Latin American group was divided into three camps. First were the relatively hard-line states that supported the 200-mile territorial sea position. These were predominantly South American, and centered on the nine adherents to the Declaration of Montevideo.[18] Second were the relatively moderate states that supported the patrimonial sea idea laid down in the Declaration of Santo Domingo. These were mostly Central American and Caribbean states, five of which had earlier joined the Montevideo states in the Lima Declaration. Their position was less firmly established than that of the Montevideo states because most of them did not have unilateral legislation establishing broad maritime claims. Third were the states not committed to any of the declarations, including the two landlocked countries and several of the Caribbean island states. Santo Domingo was the last Latin American declaration before the conference, and left the Latin American states divided, as shown in Table 8.1.

Less than a month after the Declaration of Santo Domingo, the center of regional activity shifted back to Africa. From June 20-30, 1972, representatives from 17 African states met at a seminar on the law of the sea in Yaounde, Cameroon. The seminar was also attended by a number of observers from developed states including the United States, the Soviet Union, Canada, Israel, Spain, and Switzerland. It attempted to formulate more detailed positions than those reached in the three OAU resolutions of the previous year. The seminar participants themselves did not constitute a subgroup within the African group, but contented themselves with recommending their conclusions to the OAU.

TABLE 8.1

Adherents to the Three Latin American Declarations

	Montevideo	Lima	Santo Domingo	Uncommitted	138/49
Hard-line states					
Argentina	X	X			
Brazil	X	X			
Chile	X	X			X
Ecuador	X	X			X
El Salvador	X	X			X
Panama	X	X			X
Peru	X	X			X
Uruguay	X	X			X
Nicaragua	X	X	X		
Moderate states					
Colombia		X	X		X
Dominican Republic		X	X		
Guatemala		X	X		X
Honduras		X	X		
Mexico		X	X		X
Costa Rica			X		
Haiti			X		
Trinidad and Tobago			X		X
Venezuela			X		X
Uncommitted states					
Barbados				X	
Bolivia				X	
Cuba				X	
Jamaica				X	X
Paraguay				X	
Guyana				X	X

Note: Although the Latin American states were divided on the question of national jurisdiction, the Latin American proposal on the international seabed regime had support from all subgroups except the two landlocked states.

The Yaounde seminar conclusions represented a substantial advance over the three OAU resolutions, but they still contained sufficient ambiguity to indicate that the African states were not yet close to consensus on a regional law of the sea policy. The seminar recommended a unilateralist position similar to that in the Lima Declaration, but went on from there to advocate (not unanimously) a 12-mile territorial sea and an economic zone of unspecified extent. The economic zone looked rather like a more vaguely stated version of the patrimonial sea, but with two important variations. First, the conclusions strongly suggested that the zone would include the continental shelf within a single fixed distance limit. This would in effect disallow claims made by the Latin American states, Canada, Australia, and many others to extend their seabed jurisdiction to the edge of the margin. Second, they proposed that "the exploitation of the living resources within the economic zone should be open to all African states, both landlocked and near landlocked." This, in marked contrast to the Latin American declarations, reflected the different weight of noncoastal states in the two regional groups. In addition, the seminar reinforced the strong shift in emphasis to matters of coastal jurisdiction by recommending that the African states unilaterally extend their sovereignty over an economic zone, or at least over the continental shelf, and that they take a common stand in upholding such an extension at the Law of the Sea Conference. The whole economic zone movement at the Yaounde seminar can be seen as an extension of the second OAU resolution stimulated by, among other things, the activity at the AALCC, the Group of 77 resolution, Latin American activity, and the debates in the SBC.

The seminar did not reach any detailed conclusions about the international seabed regime and authority, except that it categorically rejected any system of weighted voting or veto in the authority, and considered that the authority "must be structured and operate in such a way that the developing countries should be the primary controllers and beneficiaries." The seminar thus followed other regional precedents in giving a definite second place to the international seabed regime question.[19]

An event related to the Yaounde seminar was the submission by Kenya to the summer 1972 session of the SBC of draft articles on the exclusive economic zone concept (138/SC.II/L.10).[20] These articles were based partly on the work done earlier by Kenya in the context of the AALCC meetings (Kenya was one of the prime movers behind the Yaounde conclusions). The Kenyan paper presented a detailed and unambiguous formulation of the exclusive economic zone idea, unencumbered by the need for political compromise that had clouded the Yaounde seminar conclusions. In most respects the Kenyan formulation paralleled the Yaounde conclusions, but it made explicit the strict limitation of the zone, including the continental shelf, to a 200-mile limit, and it made clearer the allocation of extensive powers to the coastal state to control marine pollution and scientific research. As an African creation, the economic zone concept thus bore strong resemblance to the Latin American patrimonial sea idea. The major differences between the two reflected geopolitical differences between the African and South American continents, namely on the question of the continental margin beyond 200 miles and on the question of sharing with the noncoastal states within the zone. Despite the differences between them, however, the important element in both proposals was that they were alternatives to the extreme 200-mile territorial sea position. In this sense they offered grounds for compromise between the advocates of wide national jurisdiction and the advocates of narrow jurisdiction, and therefore appeared to offer most hope for reconciliation of conflicting maritime, coastal, and noncoastal interests both within and outside the Group of 77.

The year 1973 was a bit quieter in terms of regional activity, because most attention was concentrated on the last sessions of the Seabed Committee. The AALCC, which continued to discuss the law of the sea at its 14th meeting in January, served as a useful bridge between African and Asian states on law of the sea policy development. But it reached no conclusions and was generally overshadowed by events elsewhere. The Arab League, previously inactive in law of the sea matters, got as far as setting up a committee of experts on the subject, but the main weight of Arab activity remained in the African group and the Group of 77.

The only regional event of importance in 1973—and before the Law of tho Sea Conference—was the Declaration on the Issues of the Law of the Sea made by the OAU heads of state meeting in June.[21] This declaration carried considerable political weight since it represented a consensus among the 41 members of the OAU. It also represented a considerable diplomatic victory for Kenya, since the declaration supported virtually without reservation the economic zone concept as outlined by Kenya, thus making it in effect an African position. The only difficulty was that the African states were unable to agree on a 12-mile limit for the territorial sea, and so left the question open. Other than this, however, they accepted the 200-mile maximum limit for both seabed and superjacent waters, the provision for sharing of living resources with noncoastal states of the region, and the granting of jurisdiction over scientific research and marine pollution to the coastal state within its zone. With this declaration, the idea of 200 miles as the limit of national jurisdiction took an enormous step forward. It meant that a solid majority of developing countries in two strong regional groups had declared themselves in favor of some form of 200-mile limit, and that such a limit could be devised with the consent of a substantial group of landlocked and shelflocked states.[22]

The declaration was relatively non-specific regarding the international seabed regime and machinery. It reaffirmed the moratorium resolution and advocated a strong and comprehensive machinery dominated by the assembly. The machinery was to have powers to explore and exploit, to minimize adverse economic effects, to protect the marine environment, and to regulate scientific research. In general, the OAU position accorded with that outlined in the Tanzanian proposal on the seabed regime and machinery (138/33), making Tanzania the leading African spokesman on the international seabed area and Kenya the leading spokesman on the economic zone.

The OAU also commented on a number of other issues relevant to seabed politics. The African states accepted the right of archipelagic states to draw baselines around their outer islands, but expressed reservations as to whether all islands should be allowed to claim full maritime zones regardless of size and population. These two positions, while in some respects contradictory, can be explained by the success of the lobbying by Indonesia and the Philippines among the developing countries on the one hand, and by the lingering concern that the common heritage would be too much eroded by excessive national claims on the other. More important was the proposal that high seas fishing be subject to greatly increased regulation either by an international regime and machinery or by large additions to the powers of existing fisheries organizations. This was much more strongly put than the similar item in the Declaration of Santo Domingo and came close to, although it did not specifically recommend, giving powers over fishing to the international seabed regime and machinery. Any such suggestion was anathema to the maritime powers, particularly the United States and the Soviet Union, and boded ill for the negotiations on the international seabed area if the link was ever made explicit.

The high level of regional activity on the law of the sea among the developing countries during the period 1971-73 was an important accompaniment to the negotiations in the Seabed Committee. It resulted not only in the emergence of strong regional positions in Africa and Latin America, but also in a widespread acceptance among the developing countries of a 200-mile coastal economic zone covering both the seabed and the water column. The African group in particular, with its formidable unity of position, became a major actor in the SBC, and both the African and the Latin American groups had an obvious option to fall back on regional positions should the global negotiations fail. Although differences remained over questions like the continental shelf beyond 200 miles, sharing with landlocked and geographically disadvantaged states, and navigation rights within 200 miles, the conversion of the African group to a 200-mile position represented a huge gain for the original Latin American 200-mile states. With the OAU declaration in 1973, the 200-mile position became virtually impregnable, and moved to a dominant position in the negotiations.

By pointing the way to a compromise with noncoastal states, the OAU declaration further strengthened the 200-mile position by giving it a potentially more universal applicability, and therefore helped to ease the tension created with the landlocked states in the clash over the limits study in 1972. The universality of the African compromise was, however, limited by the relatively low cost of the 200-mile seabed limit and the fisheries sharing scheme for the African states (because of narrow margins and relatively limited coastal fisheries). The same solution elsewhere, for example in Latin America, would involve much greater sacrifices for some states (margin for Argentina, fisheries for Peru), and would therefore encounter proportionately greater political resistance.

In marked contrast to this extensive development of policy among the developing countries outside the SBC, very little external coordination was visible among the developed states during this period. The European Economic Community (EEC) sent an observer to the SBC in 1972 and made some efforts to begin coordinating the policies of its members, but did not really even begin the complex process of producing a unified position. The United States and the Soviet Union agreed to broaden their cooperation on oceanographic matters at the Nixon-Brezhnev summit of June 1973, but this was more the public window dressing of detente than a serious commitment. The developed states preferred to pursue the intricacies of alignment building almost wholly within the context of the SBC. This preference reflected, among other things, the tendency among the developed states to pursue detailed technical solutions rather than the sweeping legal and political ones preferred by most developing countries. Regional groups by their nature were much more suitable for the latter than for the former.

Accompanying this burst of regional activity was a continuation of the steady flow of unilateral claims. Between 1971 and 1973, 24 states made claims to the maritime resources adjacent to their coasts, and the great majority of these claims took the form of extensions of the territorial sea. Nine states, including France, joined the large group already claiming 12 miles, and one, Malta, joined the small six-mile group. Much more important, though, were the nine states that nearly doubled the size of the group claiming territorial seas of more than 12 miles. Nigeria, Mauritania, and Ghana claimed a territorial sea of 30 nautical miles. Gambia,

Tanzania, and the Malagasy Republic claimed 50 miles. Gabon claimed 100 miles, and Sierra Leone and Somalia joined the Latin American states in claiming 200 miles. All nine of these states were African, and they brought to 12 the number of African states claiming territorial seas beyond 12 miles. Their claims firmly established the territorial approach to coastal state jurisdiction on two continents, and so further strengthened the position of the original Latin American territorialist states.

Four of the nine African states made other claims additional to those to the territorial sea. Mauritania joined the many states claiming the continental shelf out to 200 meters or the limit of exploitation, while the Malagasy Republic abandoned that group and, with a claim of 150 miles, joined the growing numbers of states claiming the shelf out to a fixed distance limit. Gabon added a 50-mile fishing zone to its 100 mile territorial sea, and Ghana added a 100-mile fishing zone to its 30-mile territorial sea. South Vietnam, Iran, Morocco, Senegal, and Iceland also claimed fishing zones beyond the territorial sea, varying from 50 to 122 miles in breadth.

The concentration of large claims in Africa paralleled the rapid development of regional activity on the law of the sea in the OAU, and was a further sign of the emergence of Africa as a major center of activity in the law of the sea negotiations. The claims were a substantial breach of the restraint generally considered necessary to successful international negotiations—especially since many of them fell outside the main trend embodied in the 12-mile territorial sea and the 200-mile economic zone. The continued stress on territorial sea claims and fishing zone claims reflected both the Latin American tactic of picking the weakest area of international law, and a genuine shift of concern away from the issue of rights on the shelf toward the more complicated issue of rights in the water column.

ALIGNMENTS ON LIMITS AND RIGHTS

Unilateral claims, when added to the various regional declarations and resolutions and to the statements and formal proposals made in the Seabed Committee between 1971 and 1973, combined to give a fairly clear picture of the state of opinion on defining the boundary between national and international seabed areas. By the end of the last SBC meeting in the late summer of 1973, nearly 100 states had expressed an opinion on the boundary question, and three contending schools could be identified.

The first school consisted of those states that opposed a 200-mile limit as being too large. Most of these states supported proposals for a narrower limit, 200 meters and 40 miles being the most commonly preferred alternatives. Although there was some confusion here between seabed and water column issues, virtually all of these states mentioned the seabed, their concern being to give the maximum area to the international seabed authority. Nineteen states expressed views along these lines, 13 of them landlocked or shelflocked and 12 of them developed states both Western and Soviet.[23] Both the United States and the Soviet Union appeared to favor this point of view because of their interest in narrow coastal state limits of jurisdiction in the water column, but both also favored the opposing view of extensive coastal state jurisdiction over the continental margin. Consequently the landlocked and shelflocked states, particularly the developed among them, were the main force behind this point of view. It is important to note that no African states openly supported it despite the large number of landlocked states on that continent. The main strength of the school was its potential to attract the rest of the landlocked and shelflocked states with their substantial bloc of votes.

The second school consisted of those states that supported coastal state jurisdiction over the seabed extending out to the edge of the continental margin. Visible support for this point of view came from the United States, the Soviet Union, Canada, Italy, Australia, Iceland, Sweden, Norway, Ireland, Senegal, and 13 Latin American states—a total of 23 states, many of which were identified in Chapter 6 as members of an emergent coastal group.[24] Seventeen of these states also supported the 200-mile limit in combination with the geographical criterion, and most of them opposed a strict 200-mile limit because it would force them to sacrifice parts of the margin that they already claimed to control under the terms of the 1958 Continental Shelf Convention and the 1969 ICJ judgment on the North Sea shelf case. Canada, Norway, Argentina, and Australia were the leading exponents of this latter argument, all of them in defense of their extensive margin areas beyond 200 miles. A total of about 40 states had physical margins extending beyond 200 miles off their coasts, but more important than mere numbers was the fact that this school included many of the major coastal states whose agreement was essential for any new law of the sea convention. Fewer than a dozen states stood to lose very large areas if a choice had to be made between the margin and a flat 200-mile limit, and no state had actually made a unilateral claim to the margin, although some of the vague geographical claims could be (and the 1958 Convention was) interpreted in that sense.

The third and largest school consisted of those states favoring a flat 200-mile limit. As we have seen, this point of view was heavily related to water column issues, on which opinions differed sharply within the 200-mile context. On seabed issues, however, there was considerable uniformity, the only difference being a conceptual one relating mostly to water column issues, namely whether the coastal state had sovereign rights to the margin itself or only to the resources of the margin. Not including the 17 that also wanted the margin, 57 states supported the idea of a flat 200-mile limit to coastal state jurisdiction over the seabed. Most of these supported the 200-mile limit for an economic zone, but some supported that limit for a territorial sea, some for a patrimonial sea, and some for the continental shelf. The backbone of this school was the OAU, all but one of whose 41 members (Senegal) had accepted the organization's support of a 200-mile economic zone in its 1973 declaration. All but four of its adherents were developing countries. France, Spain, Denmark, and Malta all supported it, as did major Asian coastal states like India, Pakistan, Sri Lanka, and China.[25]

A possible fourth school relevant here consisted of those states concerned with the rights of archipelagic countries. Their position was that archipelagic states had the right to all the seabed enclosed by baselines drawn around the outside edge of the archipelago, regardless of the depth of the overlying waters. A group of four states—Fiji, Indonesia, the Philippines, and Mauritius—advocated this view, and had the potential to attract other archipelagic states in the Pacific Ocean.[26] In practice, however, not too much difference in terms of seabed would result from the application of the archipelagic principle rather than giving each island a 200-mile zone. Consequently this group made little difference to purely seabed considerations.

At the end of 1973, then, the debate on the boundary between national and international seabed areas had reached a three-way split. The flat 200-mile position had unquestionably gained the most numerical support, but was vigorously opposed by the narrow zone and margin states. The margin states faced the nearly unanimous hostility of the African group, and also of the narrow zone school, with its strong contingent of landlocked and shelflocked states. And the narrow zone position, which at one point appeared to be the major challenger to the 200-mile zone, was opposed by both the 200-mile and the margin states, and looked more and more like a futile protest against the tide of coastal state expansion.

The apparent lack of grounds for compromise among these three positions was not as serious as it might have been because of a second theme in the boundary debate that emerged strongly during 1973. Instead of concentrating on defining the boundary, as had been the case up to 1973, the discussion expanded to consider not only the limits of coastal state jurisdiction over the seabed, but also the nature of the rights that the coastal state would have within whatever boundary was set. This second theme was closely related to the limits question, and had the effect of making the discussion more complicated while at the same time opening up grounds for compromise.

The debate on rights within the limits of national jurisdiction revolved around four basic proposals, and tended to increase the interconnectedness of seabed and water column resource issues. First was the position taken by the more extreme coastal states, particularly the Latin American territorialists, that the coastal state had exclusive rights to all resources, both seabed and water column, within the limits of national jurisdiction. The noncoastal states opposed this view vigorously, and fear of it was a primary motivation for the narrow limits school of thought.

Second was the position taken by a number of landlocked and shelflocked states that they, and others in a similar position, should have equal rights to the resources on the continental shelves and in the economic zones of neighboring coastal states.[27] Proposals here varied from regional economic zones to elaborate revenue sharing schemes; by the end of 1973, this approach looked ripe to replace narrow zone proposals as the main line of the landlocked and shelflocked states.

Third was the position first adopted by the United States, Malta, and Canada in their respective proposals on the international seabed regime and machinery that coastal states should share revenue with the seabed authority from exploitation on their continental shelves. There were various proposals on how this should be organized—varying mostly in terms of the distance from shore at which sharing should begin—and a number of leading landlocked and shelflocked states adopted the idea in addition to their equal rights position.[28] This looked like the most probable avenue for compromise between the margin states and those advocating a flat 200-mile limit—that is, revenue sharing from the margin beyond 200 miles.

Fourth was the position adopted by the African states, and also by Uruguay in a much weaker form, that landlocked and shelflocked states should have the right to share in the living resources within the jurisdiction of neighboring coastal states.[29] This proposal was put forward by the most conciliatory group of coastal states, and differed from the other three in that it was an offer to trade off a position on one issue against a position on another. In effect these coastal states were offering the landlocked and shelflocked states the right to share in fishery resources in return for the latter giving up their claim to share continental shelf resources. The compromise was an important one not only because it had been accepted within the African group, but also because it offered a way around the extremely contentious issue of sharing continental shelf resources. Coastal states were exceedingly reluctant to share what most of them interpreted as being theirs already under existing international law. Hopes of offshore oil elevated this reluctance into adamant refusal, and even in Africa the idea of such sharing had been strenuously rejected by the coastal states. By the end of 1973, this position, along with a limited amount of sharing of continental shelf revenues with the international seabed authority, looked like the most fertile ground on which a global compromise might be worked out.

Underneath the slowly evolving negotiations centered on the SBC, the threat of unilateralism continued to rumble all through the period 1971-73. In addition to the unilateral claims actually made by individual states, there was a persistent undertone of unilateralism in many of the regional declarations and resolutions. On the boundary question this threat came mainly from the developing countries—quite explicitly in the Montevideo and Lima Declarations and in the Yaounde seminar conclusions, and rather more vaguely in the OAU and Group of 77 ministerial meeting resolutions of 1971. Some threats of unilateral action were also made within the SBC, for example by India, Pakistan, and Sri Lanka,[30] and there were increasing signs of unilateral thinking from some developed states like Canada, Norway, Iceland, and the United States, whose coastal fisheries were under pressure from foreign fleets.

For the most part, this unilateralist sentiment had little direct connection to seabed issues. It was nearly all oriented toward water column issues, particularly territorial sea and fishing matters, and was widely held at the time to be the major threat to the sense of accommodation that was thought necessary if the negotiations were to succeed. Although it did not relate directly to seabed issues, it did threaten the whole package of law of the sea negotiations, and was thus of the highest concern to those interested in creating an international seabed authority. Indeed, the perceived threat from unilateral action was one of the principal pressures behind the shift from the Seabed Committee to the Third UN Conference on the Law of the Sea at the end of 1973.

THE GROUPS

Just as the debate on policies had evolved over the 1971-73 period, so also had the pattern of political alignments. It has already been suggested in the discussion of both regional activity and the policy debates that the developed versus developing alignment on the international seabed regime remained relatively unchanged, that the African group grew stronger, and that the landlocked and shelflocked states were acting in unison against the expansion of coastal state jurisdiction. It is worthwhile here to review developments among the various groups active in the SBC. Since no voting took place in the SBC, the principal recorded evidence for group activity lies in the patterns of sponsorship of the many documents presented to the committee.[31]

The situation among the external groups remained much the same as it had been in 1970 except for the dramatic strengthening of the African group. The patterns of sponsorship in the SBC, however, provide some interesting insights into the strengths and weaknesses of the groups.

The Group of 77 eventually numbered nearly 60 states in the SBC, and it continued to play an important role as a negotiating forum for the developing countries. In concrete terms, however, much of its value as a policy alliance group was eroded by the addition of large numbers of issues to the agenda in 1971, by the increasingly complex and detailed level of the debate on the international seabed issue, and by the massive increase in the numbers of developing countries represented in the SBC. Even in SC.I, where the Group retained a large measure of its influence as a rallying point for the developing countries, none of the vital papers on the international regime and machinery were put forward in its name. This, however, was largely offset by a growing unity of the Group around the Latin American proposal during 1971-73, and so SC.I remained the forum in which the Group of 77 was most active and influential. In SC.II, the Group served as a major rallying point for the developing countries only on the issue of compiling a list of subjects and issues for the Law of the Sea Conference.[32] It had some success on the issue of scientific research in SC.III,[33] but was completely unable to overcome its internal divisions on crucial geographical and economic issues like the continental shelf and the economic zone. In general it was conspicuously unsuccessful in gaining the support of the landlocked and shelflocked states, but did maintain its successful absorption of the metal-exporting states.

The policy disarray within the Group of 77 was one of the key factors in the slowness with which the SBC pursued its mandate between 1971 and 1973. Because the Group had a very strong symbolic value for many developing countries, and because many of those countries saw the law of the sea as an issue between developed and developing states, the Group of 77 took on an importance that transcended its poor showing on policy. Many developing countries were reluctant to move unless they could all move together within the framework of the Group of 77. Since they could not yet agree, general progress had to await the evolution of compromises within the Group. In this sense, much of what went on in the SBC can bé interpreted in terms of the internal politics of the Group of 77.

The Latin American states unquestionably formed the most active regional group of developing countries within the SBC. By 1972, 17 Latin American states were represented on the Committee, and regardless of the state of agreement among themselves, they met as a negotiating group. On some matters, the Latin American group could be considered as a group of the whole (excluding Bolivia, the only landlocked state). This was true for issues such as the international regime and machinery, and the list of subjects and issues, where nearly all the members of the group could be found sponsoring a distinctly Latin American proposal.[34] On other matters, the pattern of sponsorship confirmed the subgroups revealed by the various regional declarations emanating from Latin America. A group of territorialist states (Ecuador, El Salvador, Peru, Uruguay, Panama, Brazil) acted together, though not always all together, on several proposals,[35] and Venezuela, Mexico and Colombia, the leading states in the Santo Domingo Declaration, also jointly sponsored a document on the territorial sea.[36]

The Latin American states cosponsored proposals within their regional group more frequently than they participated in Group of 77 proposals, and appeared to be less strongly attached to the Group of 77 than most of their African and Asian counterparts. They very seldom resorted to documents sponsored by a single state, and only seven of them showed any inclination to cosponsor with groups other than the Latin American or the 77. In other words, strong regional groups and the Group of 77 constituted the main focus of activity for most Latin American states.

The emergence of the African regional group was much more visible outside than within the SBC. Twenty-four OAU members were active within the committee, but the group preferred to formulate its policy at OAU meetings so that all of its members could participate. Within the SBC, only one distinctively African proposal was made, and that was sponsored by 14 coastal states representing less than 60 percent of the group's membership.[37] By contrast, the African states were very active in Group of 77 proposals, although like the Latin American states their landlocked contingent participated neither in the regional group nor in the Group of 77 proposals. Only three African states sponsored proposals by themselves (Kenya, Tanzania, and Zaire), but 14 cosponsored with groups other than the African or the 77.

The rather uncoordinated cosponsoring behavior of the African states in the SBC contrasted oddly with the formidable policy unity achieved in the 1973 OAU declaration. The African group was much more coherent than it had been in 1970, and its 200-mile economic zone position was one of the key proposals affecting both seabed and ocean politics. But in many respects the group was like a smaller version of the Group of 77, in that it contained a great diversity of conflicting geographical and economic interests bound together by a strong symbolic commitment to African unity. Within the context of international negotiations, the African states also retained their attachment to the Group of 77, an attachment that encouraged a low profile policy for their regional group within the SBC.

The Arab states had no visible regional group of their own, and seven out of ten of them were members of the African group. Virtually all of their cosponsoring activity was in the context of the Group of 77, and only four of them cosponsored with groups other than the African or the 77. No Arab states sponsored proposals by themselves.

The 15 Asian states within the SBC likewise had no regional group of their own, and also tended to participate heavily in Group of 77 proposals. To a markedly greater extent than other developing countries, the Asian states sponsored proposals by themselves and cosponsored proposals with groups other than the Group of 77. Despite the relatively small number of Asian states in the SBC, they participated in nonregional and non-Group of 77 proposals more times than all the other developing countries put together, and also placed more single sponsored proposals before the committee than all other developing countries taken together.

Among the developed states, the only regional group visibly active in the SBC was that centered around the Soviet Union. The eight members of the group (the Soviet Union itself counting as three) were not as conspicuously unified in action as normal within the SBC, though there was cooperation among some of them on four proposals, and they all abstained on Resolution 3067 in the 1973 General Assembly meeting.[38] Five of the members cosponsored with other groups, including Romania with the Group of 77, and Hungary and Czechoslovakia with the landlocked and shelflocked states, and the Soviet Union sponsored nine proposals by itself. Thus, although it was active as a forum for cooperation, the Soviet group was clearly feeling the pressure created by the very different economic and geographical characteristics of its members.

The Western developed states had neither an active regional group of their own nor a super-regional equivalent to the Group of 77. The EEC states still showed no sign of emerging as a group, and so by default the primary activity of the Western states was in single sponsored proposals and in proposals sponsored by nonregional groups. A dozen Western states, conscious of both their relative political isolation and their superior technical knowledge, were responsible for nearly two-thirds of the single sponsored proposals submitted to the SBC, a number amounting to over one-third of all proposals submitted during the six years of SBC meetings. These same states poured a nearly equivalent amount of energy into issue-specific group proposals, and because of their lack of any regional or political group focus, were very often the driving force behind the creation of such groups.

The United States continued to behave rather like a group in its own right, and did not once cosponsor with another state in any of the 16 proposals it placed before the committee. It did, however, consult with other maritime powers, including the Soviet Union. American policy shifted from its 1970 position in accordance with complicated internal politics among various domestic interest groups, which have been discussed in detail by other writers.[39]

Issue-specific groups, because of their more tenuous nature, linked to negotiation, were much more affected by the passage of time than their more broadly-based regional/political counterparts. During the period 1971-73, a number of the groups identified in Chapter 7 as being potential developed fully independent status, so that by the end of 1973 at least five independent issue-specific groups could be identified on the basis of patterns of cosponsorship. These included some with little bearing on seabed politics, such as a group of nonmaritime states bordering on straits, and a group of African and Asian states concerned with the presence of foreign-owned islands close off their coasts.[40] In addition there were also some ad hoc coalitions that were insufficiently long-lived to deserve the title of groups, and these also were not important to seabed politics.[41] In terms of seabed politics, the important issue-specific groups were the landlocked and shelflocked state group and the coastal state group.

The landlocked and shelflocked state group had 13 members in the SBC, but the full extent of its influence was more accurately measured by the 31 landlocked and shelflocked states that sponsored Resolution 3029B in 1972. The group was one of the most active in the committee, and as was discussed above, many of its members took similar positions on the question of national jurisdiction over the continental shelf. It presented five proposals to the SBC between 1971 and 1973, all sponsored by between six and 11 of its members, and as noted its members were seldom to be found in either regional or Group of 77 cosponsorings within the SBC.[42] The group was an expansion of the purely landlocked state group already independent in 1970, and wielded the formidable voting strength of the two major categories of "geographically disadvantaged" state in combination. Its growth was clearly a reaction to the expansion of coastal state claims and aspirations, and the rift between coastal and noncoastal states was one of the major weaknesses in the regional/political groups. In return for its growth, the landlocked and shelflocked state group paid the price of potential internal splits along three fault lines: developed versus developing, landlocked versus shelflocked, and regional interest versus group interest. An example of the latter danger was evident in the OAU declaration, in which all of the 13 African landlocked states appeared to accept a regional compromise on the economic zone question.

The coastal state group developed much more slowly than its landlocked and shelflocked counterpart, largely because of substantial differences among coastal states on what was the best policy to cover their common interests. Because of these differences, two coastal state alignments evolved during the 1971-73 period: a moderate group composed of states from all regions except Eastern Europe, and a territorialist school, still in the group formation stage, centered in Latin America.

The territorialist school has already been discussed in the context of the Latin American regional group and African unilateral claims. The moderate coastal state group (hereafter, the coastal state group) was a loose association of states that began forming in 1972 and became visible in two multi-sponsored proposals in 1973.[43] Canada played a major role in forming the group, which included all the "renegade" states from among the developed countries (Australia, Canada, Norway, Iceland, and New Zealand), the leading Latin American moderates (Colombia, Mexico, Venezuela, Chile, and Trinidad and Tobago), two of the archipelagic states (Indonesia and Mauritius), and some of the more influential African and Asian states (India, Sri Lanka, Kenya, Tanzania, Nigeria, and Senegal). The coastal state group was loose in the sense that it did not have a tightly defined policy consensus, but it was nevertheless an important focus for consultation and cooperation among most of its members. In particular, it was important for the five developed states, which unlike the developing members had no other group within which they could work. The group's apparent ambition was to develop the moderate coastal state position into a compromise policy around which the Law of the Sea Conference might coalesce. Its weakness lay in its own fragile unity and in the threat from competing groups like the territorialists and the African group.

The emergence of two coastal state alignments, plus others like the archipelagic group and the African group with coastal interests and policies, meant that the coastal state interest as a whole was fragmented along both organizational and policy lines by the end of 1973. Thus, while the coastal state movement in general was the strongest trend in the international negotiations, the political organization of coastal states was divided and weak.

In terms of seabed politics, then, the relationship between groups and policies within the SBC was as follows. In SC.I, all the outstanding disputes on the international seabed regime and machinery were divided along developed versus developing lines. The Group of 77 was the dominant group within the subcommittee despite the fact that it had not formally produced a joint position paper. Its strength was based on a widespread acceptance by the developing countries of the view put forward in the Latin American paper (138/49), and on the use of the Group by the developing countries as a rallying point in the negotiations. This use of the Group of 77 framework by the developing countries forced the continuance of the negative school of developed states, but this did not appear as coherent as it had been in the early days of the SBC. After the flurry of activity associated with the regime and machinery proposals in 1971, no other groups were visibly active in SC.I. This left the field clear for the Group of 77 to develop as the dominant player, arrayed against a loose consortium of developed states among which the United States was the most prominent. As it had always been, the confrontation here was between technological power and voting power, complicated by heavy external associations between the adversaries arising from non-law of the sea matters. An important factor in the whole policy and group configuration in SC.I was that the issue under discussion continued to be largely isolated from the general law of the sea matters under discussion in the other subcommittees.

In SC.II, as discussed in this chapter, the relationship between policy and group alignments evolved in a much more complex pattern. On the question of national jurisdiction over the seabed, the Group of 77 played almost no part at all, and the developed versus developing alignment was largely irrelevant. The key alignments here were the landlocked and shelflocked states, the territorialists, the coastal state group, the African group, and to a lesser extent the archipelagic state group.

The landlocked and shelflocked states provided the main strength behind the narrow limits position and behind the idea of rights for all states in the coastal zone. They also provided some of the strength behind the idea of sharing continental shelf revenues with the seabed authority and some of the strength (African landlocked only) behind the idea of trading off sharing in fish resources in return for renouncing rights to continental shelf resources.

The coastal state group and the territorialists were split between the margin and the flat 200-mile positions, although both leaned toward the former. Whatever the boundary, very few members of either group advocated sharing continental shelf revenues with the international authority or with geographically disadvantaged states, although the territorialists were more adamant in this than the coastal state group.

Between these two camps stood the African group, with its own regional compromise covering their dispute, and also the United States, with its interest in carefully balanced trade-offs on a whole package of issues. In contrast with SC.I, where alignments were almost exclusively along traditional economic and political lines, the alignments in SC.II on seabed issues were largely along geographical lines. Regional alignments were only important in the case of the African group, which was sufficiently strong to overcome its geographical divisions on the issue. The continental shelf issue was also much more linked with, and confused by, water column issues, than was the international seabed issue in SC.I.

THE STATE OF SEABED POLITICS AT THE END OF THE SBC MEETINGS

In spite of the appearance of great confusion and conflict on the continental shelf issue in SC.II, seabed issues there were in many respects in a healthier condition than seabed issues in SC.I. Much less destructive polarization developed on the shelf than on the international seabed regime, and the paths to possible compromise on the shelf were more clearly marked than those in SC.I. There was a great deal of genuine movement in the development of policy on the shelf, and although some of this was attributable to the earlier stage of the negotiations, it contrasted favorably with the static position-building in SC.I.

But although the situation on the shelf issue taken by itself looked quite promising, the situation on the larger coastal zone issue, with which the shelf issue was inextricably linked in SC.II, looked bleak. Very deep differences of opinion existed on many of the water column issues in the zone, and progress in SC.II as a whole had been so disastrously slow all through the period 1971-73 that the subcommittee's work was much further behind than even SC.I's in terms of its suitability as preparation for a conference. The subcommittee had spent 1971 completing a general discussion and had made a start on the first task of its mandate: drawing up a list of subjects and issues. In 1972, it undertook more detailed discussion, managed to complete the list of subjects and issues, and even made a small start on considering draft articles. During this year, the idea of the economic zone gained much ground as the centerpiece of the subcommittee's work. In 1973, having completed the list, the subcommittee spent almost all its time receiving the flood of draft articles, working papers, and other proposals arising from the completion of the list.

Compared with even the single volume of alternate texts that came out of SC.I at the end of 1973, the four volumes of SC.II's papers were a hopeless mess, almost useless as a basis for serious negotiations. A start had been made at putting the mass of proposals into comparative tables and consolidated texts, but this represented the beginning, not the end, of a proper preparatory process.

Much of what went on in SC.II was not relevant to the subject of this study, and so it is not necessary here to trace the course of SC.II in as much detail as was done for SC.I. The slowness of SC.II's progress had some of the same causes as that of SC.I. In particular, SC.II was afflicted by the enormous number and complexity of the issues before it and by the many deep differences of policy involved in some of those issues. The result of slowness in SC.II, however, was very different from its result in SC.I, largely because the initiative at stake in SC.II belonged to coastal states, whereas the initiative at stake in SC.I belonged to international organization. Slowness of progress in SC.I tended to weaken the initiative, but slowness in SC.II tended to strengthen the hand of the coastal states, and especially the more extreme among them, by allowing them more time to mobilize allies, entrench unilateral positions, and wring concessions from states in a less powerful position. Since a lot of the issues in SC.II were considered urgent by many states, slowness there, as in SC.I, created an incentive to move the discussions into a conference format where pressure would be higher to reach agreement than it was in the relatively leisurely forum of the SBC.

Despite the slowness and difficulty of the negotiations in SC.II, on the eve of the conference coastal zone issues, including the continental shelf, were unquestionably the central component of the agenda. They had replaced the international seabed regime and machinery issues as the prime focus in less than three years, despite the three-year head start of the latter in the old Seabed Committee.

Throughout the period 1971-73 the SBC worked under pressure from the 1973 deadline established by Resolution 2750C in 1970. Given the state of the committee's work by the end of 1973, however, this pressure does not appear to have made much headway against the numerous factors working to slow it down. Nevertheless, despite the increasingly obvious inadequacy of the SBC's preparatory work, the 1973 deadline itself did not become an important issue. Resolution 3029A of December 18, 1972, passed unanimously, simply stretched the deadline a bit by allowing for another full year of SBC meetings in 1973 and scheduling only a short organizational session of the Law of the Sea Conference for December 1973. Under this arrangement, the first substantive session of the conference would not be held until the early summer of 1974, and there were already hints of a further session in 1975. At the end of the SBC's 1973 meetings, the General Assembly confirmed and clarified this schedule in Resolution 3067 (December 16, 1973), passed by 117 to 0 to 10 with the Soviet group, Brazil, and Portugal abstaining. This resolution set the organizational session from December 3 to 14 in New York, with the first substantive session to last ten weeks from June 20 to August 29 in Caracas, Venezuela. Any subsequent sessions were to convene "not later than 1975", thus suggesting to the conference a deadline of two years in which to turn the enormous pile of conflicting SBC documents into "a convention dealing with all matters relating to the law of the sea."

NOTES

1. The Times, January 25 and March 7, 1974; and UN Doc. A/AC.138/90, pp. 5-8.
2. U.S. Senate, Mineral Resources of the Deep Seabed, 93rd Congress, first session on S.1134, p. 719.
3. Ibid., p. 720; and UN Doc. A/AC.138/87, p. 9.
4. The Times, February 26, 1974. This item is an excellent example of the commercial implications of scientific research that were causing such difficulties in SC.III.
5. GAOR XXVII, Supp. 21, A/8721, 1972 (Report of the SBC), pp. 12-13; A/AC.138/SR.84, p. 84, and SR.85, p. 94; and A/AC.138/SC.I/SR.47, p. 230.
6. A/AC.138/SR.85, p. 85.
7. As of 1970, the AALCC had 13 members: Sri Lanka, Ghana, India, Indonesia, Iraq, Pakistan, Japan, Egypt, Jordan, Sierra Leone, Thailand, Burma, and Nigeria. By 1973 it had expanded to include 20 states with the addition of the Philippines, Nepal, Syria, Iran, Kenya, Malaysia, and Kuwait. As this list indicates, the committee was predominantly Asian (11), with Arab (5) and African (4) states in the minority. The presence of Japan made it a mixed group, but the committee was clearly dominated by developing countries.
8. On the AALCC for 1971, see A/AC.138/34 (April 30, 1971), pp. 7-13; A/CN.4/248; and ST./LEG./SER.B/16, pp. 594-98.
9. F. X. Njenga, "Third World Countries and the Law of the Sea," address before the Financial Times Conference, February 27-28, 1974.

10. The resolutions are reprinted in Oda, The International Law of the Ocean Development (Leiden: Sijthoff, 1972), pp. 362-64.

11. Ibid., pp. 364-66

12. This was drafted by C. W. Pinto, chairman of the working group in SC.I of the SBC, and is reprinted in ibid., pp. 305-26. It was a detailed proposal on a par with the 11 submitted to the SBC in 1970-71, and advocated a strong, assembly-dominated machinery with mixed licensing/operating powers and the power to control prices on seabed mineral production from the international area.

13. A/CN.4/262 (May 1972), pp. 4-6.

14. The vagueness of this latter qualification somewhat undermined the distinction between patrimonial and territorial seas, but nonetheless the proposal clearly represented an effort to compromise from a 200-mile territorial sea position. The text of the declaration is reprinted in A/AC.138/80 and in Lay et al., New Directions in the Law of the Sea (Dobbs Ferry: Oceana, 1973), pp. 247-49.

15. Panama and El Salvador both claimed 200-mile territorial seas, while Jamaica argued that a 200-mile zone would be of no advantage since it would leave the country hemmed in by the limits of surrounding states (A/AC.138/SR.79).

16. A/AC.138/SR.80.

17. A/AC.138/SR.78.

18. These included Chile, Ecuador, and Peru, the three signatories of the Declaration of Santiago. They maintained their alliance throughout, and celebrated the 20th anniversary of their declaration in the SBC by issuing a joint communique (A/AC.138/SR.89). It is interesting to note that the interpretation given to the declaration in this communique looked much more like a territorial sea position that the more moderate interpretations given at the 1958 Law of the Sea Conference.

19. The Yaounde conclusions are reprinted in Lay et al., op. cit., pp. 250-53, and in A/AC.138/79. See also A/AC.138/SR.78, pp. 15-16; and Douglas M. Johnston and Edgar Gold, "The Economic Zone in the Law of the Sea," Law of the Sea Institute Occasional Paper no. 17 (Kingston: University of Rhode Island Press, 1973).

20. Reprinted in GAOR XXVII, Suppl. 21, A/8721, 1972 (Report of the SBC), pp. 180-82.

21. Reprinted in A/AC.138/89, and in GAOR XXVIII, Suppl. 21, A/9021, 1973 (Report of the SBC), Vol. II, pp. 4-8.

22. Uganda and Zambia later dissented from the OAU position by proposing regional instead of national economic zones (A/AC.138/SC.II/L.41, July 16, 1973) despite the fact that this position had been rejected by the OAU (A/AC.138/SC.II/SR.70, pp. 9-10).

23. Statements to the SBC, and A/AC.138/55.

24. Statements to the SBC; A/AC.138/43, and A/AC.138/SC.II/L.21, L.26 and L.36; and the Declaration of Santo Domingo.

25. Statements to the SBC; national claims (see Tables 6.1 and 6.2); A/AC.138/SC.II/L.10 and L.40; and the 1973 OAU declaration.

26. A/AC.138/SC.II/L.15 and L.48.

27. The Netherlands, A/AC.138/SC.II/L.59; Jamaica, A/AC.138/SC.II/L.55; Afghanistan, Austria, Belgium, Bolivia, Nepal, and Singapore, A/AC.138/SC.II/L.39; and Uganda and Zambia, A/AC.138/SC.II/L.41.

28. Ibid., L.39.

29. Uruguay, A/AC.138/SC.II/L.24; 14 African coastal states, A/AC.138/SC.II/L.40; the Yaounde seminar conclusions; and the 1973 OAU declaration. Ecuador, Panama, and Peru in A/AC.138/SC.II/L.27, and China in A/AC.138/SC.II/L.34 made much more vague proposals along these lines.

30. A/AC.138/SR.47, p. 2; SR.48, p. 3; and SR.54, p. 13.

31. Approximately 150 formal documents were submitted to the SBC between 1968 and 1973, the great majority during the period 1971-73, of which about one-third were sponsored by two or more states and two-thirds by states acting alone. The method of using cosponsorship patterns to identify negotiating groups is limited by the fact that often only some of the members of the group joined in sponsoring a document evolved within the group's discussions. Consequently the full membership of a group is often not identified by cosponsored documents.

32. A/AC.138/58 and 66.

33. A/AC.138/SC.III/L.55.

34. A/AC.138/49 and 56.

35. A/AC.138/SC.II/L.27 and L.54; and A/AC.138/SC.III/L.47.

36. A/AC.138/SC.II/L.21.

37. A/AC.138/SC.II/L.40.

38. A/AC.138/3 and 45; and A/AC.138/SC.III/L.23 and L.31.

39. Ann L. Hollick, "Bureaucrats at Sea," in A. L. Hollick and R. E. Osgood, New Era of Ocean Politics (Baltimore: Johns Hopkins University Press, 1974); and Ann L. Hollick, "U.S. Ocean Politics," San Diego Law Review 10, no. 3 (May 1973), pp. 467-501. David A. Kay, "Intergovernmental Dimensions of International Conferences," in Christy et al., Law of the Sea: Caracas and Beyond (Cambridge: Ballinger, 1975), pp. 96-97, makes the interesting argument that bureaucratic politics within a large state like the United States results in a preconference position so rigid that the state's ability to maneuver and bargain internationally is severely restricted.

40. A/AC.138/SC.II/L.18; and A/AC.138/SC.II/L.31, L.32, L.33, and L.43.

41. A/AC.138/SC.III/L.22 against nuclear testing in the Pacific; and A/AC.138/SC.III/L.25 on marine pollution.

42. A/AC.138/55, 72, 81 and 93; and A/AC.138/SC.II/L.39. Two African landlocked states also sponsored A/AC.138/SC.II/L.41.

43. A/AC.138/SC.II/L.38; and A/AC.138/SC.III/L.56.

CHAPTER

9

THE THIRD UN CONFERENCE ON THE LAW OF THE SEA, 1973-74, PART I: CARACAS

This chapter and the one following carry the narrative into the latest phase of seabed politics, in which activity is centered on the Third UN Conference on the Law of the Sea. The conference process has been much slower than many observers had predicted, and as a consequence this phase is likely to last until at least 1977. The narrative of this book ends in autumn of 1975, and the approach taken has been to describe and analyze the two substantive sessions of the conference using the framework developed in Chapters 6 through 8. The main conclusions on the conference have been integrated into the final chapter.

DEVELOPMENTS PRECEDING THE CARACAS SESSION

The period between the end of the Seabed Committee in August 1973 and the beginning of the Caracas session of the conference in June 1974 was a time of activity on several fronts relating to seabed politics. This activity included the oil crisis triggered by the Arab-Israeli war of October 1973, and the subsequent moves by other commodity exporters to strengthen their position vis-a-vis the industrialized states; the flare-up of conflicts over continental shelf rights in the Aegean and South China Seas; the changing of attitudes in countries like Britain in the light of ocean resource developments; the meeting on law of the sea matters of various important groups like the OAU, the EEC states, and the landlocked and geographically disadvantaged states; and the preparatory activity for the Caracas session, particularly the organizational session held in New York in

December 1973. Some of this activity was stimulated by the imminence of the conference. All of it helped to determine the conditions and the atmosphere in which the negotiations would take place.

The oil crisis, and the general move toward a developed versus developing confrontation over commodities, had a substantial impact on seabed politics at a crucial stage in the international negotiations. The unexpected use of the oil weapon by the mostly Arab members of the Organization of Petroleum Exporting Countries (OPEC) provided a model case that deeply affected attitudes and behaviors among both developing and developed states. Not only had the OPEC states successfully used commodity power against the Western states on a political level, but they had also brought about a massive and long-lasting increase in the price of their commodity and managed to coordinate policy on production controls. Their action struck a hard blow at the wasteful and self-indulgent economies of the West, and despite the fact that it also caused severe hardship among the oil-importing developing countries, it won nearly unanimous support from the members of the Group of 77 as a major step toward achievement of the Group's objectives. Among the market economy states of the West, with their heavy political dependence on maintenance of economic growth and high levels of consumption, the OPEC action created a profound disquiet about security of access to resources, and precipitated energetic moves to find technological alternatives to existing resource dependencies.

Other developing country commodity producers took inspiration from OPEC's staggering success, and this led to increased aggressiveness among established producer organizations, such as the council of copper exporting countries (CIPEC), and to moves toward formation of organizations among other groups of producer states. Jamaica (bauxite) and Morocco (phosphate) managed successful large price rises because of their position as single dominant producers. Iron ore- and bauxite-exporting states held meetings in February and March, and the CIPEC states discussed control of copper prices and possible expansion of membership in June. Similar activity could be observed among some food exporters, notably those concerned with coffee, cocoa, and sugar.[1]

Perhaps the main follow-up to the OPEC campaign came in April 1974 with the United Nations special session on raw materials and development. The special session provided ground for the Group of 77 as a whole to take up the commodity producer initiative, and led by Algeria, it used the UN forum to make a case for employing commodity power to negotiate or force a more equitable economic relationship between developing and industrialized states. Nationalization and control over price-fixing machinery by producers were major themes of the special session. At the end of it the developing countries passed a declaration and program on the establishment of a new economic order.

This flurry of action by developing commodity-exporting states had both direct and indirect effects on the commodity questions that underlay seabed politics. The gigantic rise in the price of oil precipitated by the OPEC states removed many of the cost barriers to offshore production in deeper, more hazardous waters. At the same time, the actual and threatened cuts in supply from the oil-exporting states created strong incentives among the importing states to expand production from "safe" areas off their own coasts. Thus, the already booming offshore industry received even more stimulus to continue expanding, and coastal states, both developed and developing, faced increased incentives to keep or bring under their control as much offshore oil, and therefore as much continental margin, as possible.

One of the more significant effects of this process was to give increased weight to coastal factors within those states possessing both coastal and maritime characteristics. This was perhaps most clearly visible in Britain, where a declining but strongly rooted maritime tradition competed with the demands of newly discovered oil and gas deposits under the North Sea, and also with the fact that foreign fishing in British waters was becoming more of a problem than British fishing in foreign waters. The United States, and to a lesser extent France, also felt these pressures, and so tended to move in the direction of Canada, Australia, Norway, Iceland, and New Zealand—the nonmaritime developed states that had led the way in formulating relatively moderate coastal policies.

The effect of what might be called "commodity politics" on the international seabed regime question was less direct, but no less significant, than its effect on the continental shelf question. Although the metal-exporting states had far less potential power than the oil exporters, and were also much less well organized, fear of commodity politics as a long-term trend shaped the thinking of most importing states, and resulted in increased attention being diverted to the problem of alternative sources of supply. The deep seabed, and particularly the manganese nodules, figures prominently as an alternative source, and this strengthened the hand of the mining industry and stiffened the attitude of key governments, especially that of the United States. The result was to make the United States and other potential deep sea mining states even less inclined than they had been before to accept a seabed regime that placed powers of production control in the hands of a developing country majority. In this sense, the campaign of the commodity producing states substantially hardened the conflict between developed and developing countries over the international seabed regime and machinery. At the same time, however, it rekindled interest in the work of SC.I, which had been tending to languish in the shadow of the economic zone debate.

Heightened sensitivity over oil was a factor in triggering off the long-brewing disputes over continental shelf jurisdiction in the Aegean and South China Seas. In Asia, the dispute started when South Vietnam declared its "incorporation" of the Spratly Islands in September 1973. Part of the Spratlys' significance lay in their enclosure of a large but rather deep area of potentially oil-rich shelf in the South China Sea, and in the fact that China, Taiwan, and the Philippines also claimed some or all of the islands. Early in 1974, China rejected the South Vietnamese claim and launched a military attack on South Vietnamese forces in the Paracel Islands, another disputed group north of the Spratly Islands. The Chinese forces successfully evicted the South Vietnamese, leaving the future of the Spratly group in a state of uncertainty.

This brief conflict made clear the unsettled condition of nearly all the continental shelf boundaries in East Asia, and underlined the quite dramatic rise of China as an oil-producing and exporting state on the brink of developing its own offshore oil industry. In a more general way, the conflict also emphasized the potential for open conflict over the continental shelf in many other places.

A similar, if less dramatic, conflict erupted in the Aegean following the discovery of offshore oil by the American Oceanic Exploration Company early in 1974. The discovery raised the temperature of an already simmering dispute between Greece and Turkey, in which the Turks claimed the shelf beyond the Greek islands immediately adjacent to the Anatolian coast. The Turks had even gone so far as to issue exploration permits on the shelf west of the Greek islands, thereby inflaming relations between the two traditional enemies. Tensions escalated in a series of diplomatic notes accompanied by much hostile comment in the press of both sides. Turkey made it clear that it wanted to divide the Aegean shelf on a median line between the two mainlands, leaving the Greek islands with only a territorial sea. Greece, a late signatory of the 1958 Convention on the Continental Shelf in 1972, maintained its position that islands were entitled to a full shelf and that the Aegean was basically a Greek sea. The dispute stretched into the summer, and was eventually caught up in the larger Greek-Turkish conflict over Cyprus, which resulted in the Turkish partition of that island and the fall of the Greek military dictatorship.

This conflict, and other less visible ones between El Salvador and Honduras, and Bolivia and Chile, were all stimulated by the increased concern over oil created by OPEC. These situations put added pressure on the conference both in the form of raising highly contentious issues like foreign offshore islands, and in the form of general pressure for an early conclusion and establishment of a stable legal framework within which to settle disputes. Only one delimitation agreement lightened the conference's work in this period: an agreement on a common zone for joint exploitation between Saudi Arabia and the Sudan in the contested area between them in the Red Sea.

The approach of the conference stimulated last-minute activity among some of the regional and issue-specific groups. Under the prodding of the European Commission, the nine EEC states held meetings early in the year aimed at coordinating their policies at the conference. While these meetings did not produce a document of the type popular among the developing countries, they did mark the beginning of the EEC's role as a regional group in the law of the sea negotiations.

The OAU meeting in June 1974 reaffirmed the organization's 1973 declaration in its entirety, and added to it a somewhat stronger wording covering the rights of landlocked states to access to the sea in general, and to the fish resources of the economic zone in particular. This was a significant reaffirmation of the African regional compromise between landlocked and coastal states, but the June meeting as a whole was marred by severe splits between Arab Africans and black Africans on more general issues of policy. These splits marked the beginning of a more independent and assertive role for the Arab states as a regional group within the law of the sea negotiations.

The most important of the preconference group meetings, however, was that held by 19 developing landlocked and geographically disadvantaged states in Kampala, Uganda, during March. This meeting produced the Kampala Declaration containing nine principles relating to the rights and interests of developing landlocked and geographically disadvantaged states.[2] Most of the principles related to rights of access to the sea in general, but the last four were directly relevant to seabed politics.

With regard to the international seabed regime, the Kampala states made the explicitly political demand that landlocked and geographically disadvantaged states be proportionately represented in all the organs of the international seabed machinery. They also wanted the international seabed area to "be so extensive and contain such resources, as to ensure viable economic exploitation," a statement which implied continued resistance to extensive coastal state shelf claims that would exclude nearly all oil areas from the common heritage.

With regard to the continental shelf and the economic zone, the Kampala states took a position similar to that of the ten landlocked and geographically disadvantaged states in the SBC that had demanded equal rights for noncoastal states to the resources of the zone and the shelf.[3] They called for maintenance of all existing rights that noncoastal states had under international law to resources beyond the territorial sea, and for equal rights for such states in any new jurisdiction beyond the territorial sea that might be established by the Law of the Sea Conference. These demands indicated a substantial shift toward sharing of resources within the zone and shelf areas as the major position of the landlocked and geographically disadvantaged state group, and a corresponding decline in emphasis on the width of coastal state jurisdiction. While the Kampala Declaration looked like a gain for the group in terms of common policy, the fact that the Kampala meeting was exclusively among developing countries boded ill for the overall unity of the group along developed/developing lines.

The organizational session of the Law of the Sea Conference sat from December 3 to 15, 1973. Its main tasks were to elect officers for the substantive sessions of the conference and to determine the rules of procedure by which those sessions would operate. The session was made necessary by the transfer from the Seabed Committee to the conference, because although the basic structures of the former were carried over to the latter when the SBC closed down, neither the SBC's consensus rule nor its officers automatically accompanied the transfer. The December session was vitally significant to seabed politics, because its decision on how the conference was to make decisions would be a major factor in determining the power of the various contending alignments. A requirement of very large majorities would give blocking power to relatively small groups like the maritime states, the landlocked states, and the territorialists. A requirement of simple or two-thirds majorities, on the other hand, would give power to large groups like the Group of 77. Given the known differences among states on the question of procedure, the New York session was also seen at the time as an indication of how heated the later substantive disputes would be.

In the event, the New York session turned out to be a very unpromising start. Much of the session was spent working out disputes among the Western states concerning the election of officers, and the time remaining after these were settled was insufficient to solve the even thornier problems of the rules of procedure. The central issue of procedure was how to operationalize the gentleman's agreement approved by the General Assembly on November 16, 1973, the key part of which read: "The Conference should make every effort to reach agreement on substantive matters by way of consensus and there should be no voting on such matters until all efforts at consensus have been exhausted."[4] Explicit rules had to be found to determine when all efforts at consensus had been exhausted, and what kind of vote should take place if no consensus could be reached.

The leading developed states—Britain, France, the Soviet Union, the United States, and Japan—conducted a frantic campaign to maintain an interpretation of the gentleman's agreement that would blunt the voting majority of the developing countries and allow minorities to defend their interests. They wanted the procedures to provide for maximum delay before any voting could take place, and for large majorities if voting became necessary.[5] Japan led an attack on the interim use of standard UN rules of procedure by the session, and so managed to upset even the existing basis on which the issue could be discussed. In the face of pressure from the developing countries to push the procedures question to a vote, several leading developed states made thinly veiled threats to walk out of the conference, and only with the greatest reluctance did these states concede to the use of voting to break the deadlock over election of officers.

The adamant stand of the developed great powers resulted in a deadlock over procedures right up to the end of the New York session, and opened up the prospect of the conference failing on the procedural issue before it even began to discuss the law of the sea. Fears were growing that a continuation of the procedure argument at Caracas would paralyze the conference and waste what was supposed to be time allocated to matters of substance. At the eleventh hour, however, a compromise was arranged on the basis of a Canadian proposal. This involved informal consultations on the procedure issue during the intersession period and the use of the first week of the Caracas meeting for final formal discussion. Britain, France, and the other powers agreed that the UN rules of procedure would apply until the conference adopted its own rules, and that if the rules had not been adopted by consensus before the end of the first week at Caracas, they could be adopted by vote.

The New York session ended on this indecisive note, demonstrating once again the impotence of a large block of voting power when arrayed against the holders, however few in number, of the real power to act in relation to the issue at hand. The conference would be a useless exercise without the participation of the leading developed states, and this fact enabled those states to exert strong pressure on the other participants to make concessions on this vital issue. Nevertheless, since the relatively dire threat of nonparticipation was not a weapon that could be used frequently, its use on this occasion demonstrated the importance which the leading developed states attached to this issue.

No visible accommodation on procedures emerged from the intersession discussions, but despite this the prevailing mood at

Caracas was wholly different from that at the New York meeting. In recognition of the fact that a rapid agreement on rules of procedure was essential if the conference was to have any chance of success, nearly all delegations adopted an attitude of compromise. As a result the conference was able to adopt by consensus its own rules of procedure before the expiration of its one-week deadline. Some states, such as the Soviet Union, Peru, and Senegal, which had taken extreme positions on either side of the issue, complained about the number and degree of concessions they had had to make, but the achievement of agreement through compromise nonetheless set a buoyant and optimistic tone for the substantive debates which followed.

The rules of procedure adopted were, on the whole, favorable to the protection of minority group interests. They gave substantial powers of delay to groups as small as 15 states, and virtually endless powers of delay to any group able to muster a simple majority in a vote. In combination with the gentleman's agreement and with the general recognition that a convention would need to be very widely supported, they made it likely that the work of the conference would be conducted largely by consensus. In order to decide that all efforts to reach consensus had failed, a two-thirds majority of representatives present and voting was necessary, provided that that majority amounted to at least a majority of the states participating in the session. This formula provided another opportunity for groups with less than a majority to delay the process, especially given the fact that states abstaining were counted as not voting. Should a matter of substance actually come to voting, the same two-thirds of voting/one-half-plus-one of participating formula would apply.

The adoption of these rules effectively blunted the voting power of the developing countries and so justified the exertions of the leading developed states. The rules assured that the conference, like the SBC, would proceed with agonizing slowness while it labored to accommodate the interests of many divergent groups of states. They enshrined the decision to continue to seek the broadest possible level of agreement, but did so at the cost of maintaining two dangers: first, that the conference would be too slow to reach agreement, and second, that the agreements it did reach would be too vague to provide a solid foundation for a new law of the sea.

THE CARACAS SESSION

The Caracas session did not turn out to be what most observers had expected. The Americans, among others, had made it clear that they expected the transfer into the conference format to result in an appreciable speeding up of the pace of negotiations.[6] Although nobody expected the session to produce as much as a complete draft convention, there was a widespread hope, particularly among delegations from developed states, that agreed draft articles could be negotiated on at least some of the key issues facing the delegates. In fact, no agreed draft articles at all came out of the session, which spent most of its time repairing the deficiencies in the SBC's preparatory work. Despite the failure to produce any conference-wide agreement on matters of substance, most delegates left the session with feelings of satisfaction with the work done and optimism about the next session scheduled for Geneva, March to May 1975. In other words, expectations were adjusted to match what proved possible in practice. Previous expectations were dismissed as having been too optimistic in the light of such factors as the number of states and issues, the consensus procedure, and the unfinished work of the SBC, all of which tended to slow things down.

In addition to the agreement on rules of procedure, the Caracas session accomplished a number of other tasks, all of which were essential to further progress on matters of substance. First, it allowed the nearly 50 states that had not been members of the SBC to enter into the negotiating process. This involved several weeks of general statements by those states, and accounted for much of the time used at Caracas. Second, it produced better definitions of some of the issues at stake, thereby improving understanding of exactly what had to be negotiated. This was particularly noticeable in Committee I (C.I), which had taken over the international seabed regime and machinery mandate from SC.I of the SBC. Third, it provided scope for a fuller revelation of alignments on many of the key issues, and this process resulted in some reduction of alternatives where one position attracted nearly all support at the expense of its rivals. This was true for the 200-mile limit for the economic zone (though not for the continental shelf), and constituted one of the major achievements of the session. Fourth, the session allowed Committee II (C.II) to make considerable headway in sorting out the chaos of documents bequeathed to it by SC.II of the SBC. By the end of the session, C.II had managed to organize all the proposals from the SBC and all those submitted at Caracas into a coherent body of alternative texts, and had therefore reached a position where negotiation could begin in earnest.

The nature of events at Caracas meant that seabed politics was carried on there in much the same way as it had been during the last three years of the SBC. Since there was very little hard negotiating at Caracas, positions on particular issues did not get jumbled together in bargaining and concession-making aimed at a package deal on the law of the sea as a whole. The international seabed regime and machinery debate still remained largely separate from what was going on in C.II, except, of course, for its links with the boundary question embodied in the issue of continental shelf limits. In C.II, the shelf issue remained entangled with the economic zone, but was still sufficiently independent of it to both justify and make possible separate consideration. Caracas also provided opportunity for further developments among the various alignments.

Continental Shelf Questions

The debate on the continental shelf at Caracas was simply an extension of the issues and themes established in the SBC. Discussion centered on the same alternatives for limits and rights within limits, but at Caracas alignments on these alternatives became much more clearly defined than they had been in the SBC. The effect of this clarification was to reduce the number of politically credible alternative policies. As a result, by the end of the Caracas session the shape of a final agreement on the continental shelf was quite clear, even though a good deal of bargaining and compromising remained before it achieved political acceptability.

On the question of limits, the school that opposed a 200-mile limit suffered severe loss of support, while the margin and flat 200-mile limit schools both gained adherents from the influx of new states into the negotiations. Up to the end of 1973, 19 states had declared their opposition to the 200-mile limit, and many of them had supported proposals for a narrower limit that would give more resources to the international seabed regime. During the Caracas session nine of these states shifted their support to the margin or flat 200-mile schools, while only two new states joined the opposition to wide national jurisdiction. As a result, this school was left with only 12 supporters, all of which were landlocked or shelflocked states.[7] No new proposal covering the position of this school was put forward at Caracas, and by the end of the session it had effectively ceased to represent a realistic alternative on the question of limits.

The margin school, which included 23 states by the end of 1973, more than doubled in size during the Caracas session. A few states changed over to the flat 200-mile position, but many new participants and some previously uncommitted states used the Caracas meeting to announce their support for the margin limit. By the end of the session, 48 states had declared in favor of the margin limit, nearly all of them in combination with a 200-mile limit. Notable additions to the school included Britain, India, Indonesia, and Yugoslavia, and it retained its predominantly coastal state flavor. The school was well balanced regionally except that it included only five African members, but it was divided in the sense that eight of its members supported the margin limit only on condition that there be revenue sharing with the international seabed authority.[8] No landlocked states and very few shelflocked states supported the margin position, and 23 states specifically opposed it, including six landlocked and seven shelflocked states, as well as important coastal states like Italy, Japan, Malta, Kenya, Tanzania, and Liberia. Despite this opposition, however, the margin states were in a strong position by the end of the Caracas session. Their numbers and their unilateral option ensured them a large voice in the final settlement, provided that compromises could be worked out on sharing and that some of the opposition could be set to rest by agreement on a precise definition for determining the edge of the margin.

The flat 200-mile limit school remained the largest despite losing more than a dozen members to the margin position. Twenty-three states joined the school at Caracas, and more than made up for the losses by bringing the total membership up to 63 states compared with 57 states at the end of 1973. African states still accounted for a majority (34) of the school's members, and although some major coastal states had gone over to the margin position, Sri Lanka, Malta, China, France, Italy, Japan, and the Philippines still adhered to a flat 200-mile limit. Most significant, however, was the weight of landlocked states within the school. In addition to the 13 African landlocked states, Austria, Switzerland, Bolivia, Paraguay, and Mongolia all joined the school, as did some important shelflocked states like East Germany, Cambodia, and Singapore. These states amounted to one-third of the school's members, and this indicated a substantial shift among the landlocked and shelflocked states away from the battle over limits and toward the battle over rights within limits.

The link between continental shelf and water column issues remained as strong at Caracas as it had been in the SBC, and states were still divided along territorial sea/patrimonial sea/economic zone lines. One of the main features of the Caracas session, however, was the nearly unanimous agreement on 200 miles as the limit of national jurisdiction in the water column. Although divisions remained deep on the nature of rights within that limit, even countries like the United States, the Soviet Union, and Britain came around to 200 miles as the basic framework for discussion. As a result of this, the 200-mile limit acquired an irreversible momentum.

This meant that the question of the boundary between national and international areas was much closer to solution than it had ever been before. The Secretary-General's report A/AC.138/87 of 1973 had made clear that in terms of seabed resources there was not much difference between 200 miles and the edge of the margin. In either case the international seabed authority would be left with very little oil, almost none of which would be suitable for exploitation in the near- or middle-term future. Because of this fact, the decline of the narrow limits school meant that the boundary question virtually ceased to be a problem insofar as the debate in C.I was concerned. Any settlement within 200-mile/margin limits meant that the international seabed authority would be predominantly a nodule mining operation in the foreseeable future, regardless of whether or not it received revenue from oil operations within national jurisdiction.

With the reduction of the limits issue to a confrontation between two solidly entrenched schools of thought, the question of rights within the limits took on even greater importance at Caracas than it had had in the SBC. It became very clear that the solution to the limits problem could only be found in adjustments and compromises on the matter of rights.

Predictably, the major initiative here came from the landlocked and geographically disadvantaged states, many of which had abandoned their fruitless fight against the 200-mile limit. On August 5 a group of 15 landlocked and six shelflocked states submitted a proposal designed to break the coastal state monopoly in the economic zone.[9] The sponsoring group, which contained a good balance of states from all regions except Eastern Europe, demanded that landlocked and other geographically disadvantaged states have the right to participate in the exploration and exploitation of the living and nonliving resources of the zones of neighboring coastal states "on an equal and nondiscriminatory basis." In addition, they demanded that "all states deriving revenue from the exploitation of the non-living resources of the —— zone shall make contributions to the international authority at the rate of —— percent of the net revenues." This document made the landlocked and geographically disadvantaged state group the leading exponent of both these proposals for sharing the resources of the zone.

Reactions to the proposal fell into three categories. On the question of sharing continental shelf revenues with the international seabed authority, only the United States held a similar position.[10] Neither Canada nor Malta referred to their SBC proposals along these lines, and in general the idea of sharing with the authority was not much discussed or developed at Caracas.

On the question of coastal states sharing nonliving continental shelf resources with noncoastal states, reaction to the proposal was mostly adverse. Only Haiti supported it, and a dozen coastal states including Romania, Argentina, Uruguay, Ghana, Kenya, Nigeria, India, Iran, and Pakistan went out of their way to denounce it. The great majority of other coastal states quietly ignored the proposal by continuing to refer to "exclusive" or "sovereign" coastal state rights over continental shelf resources.

On the question of coastal states sharing living resources of the zone with noncoastal states, however, reaction was remarkably favorable. This part of the proposal was a copy of the African compromise first presented in 1973. At Caracas, states from all regions spoke in favor of fisheries sharing, and so expanded the proposal well beyond the African context. By the end of the session over 100 states had supported the idea (though there were considerable differences in the degree and specificity of commitment expressed) and none had openly opposed it. Some states were notably quiet about the proposal, including Brazil, the Philippines, Japan, Britain, France, Italy, and Australia, but most other major coastal states favored some version of it.

On the basis of the alignments on limits and rights at Caracas, the outline of the likely final settlement on the continental shelf was quite clearly developed by the end of the session. Both the flat 200-mile and the margin limit schools had sufficient strength so that neither was likely to abandon the field. The difference between them appeared amenable to settlement on the basis of revenue sharing from the margin beyond 200 miles with the international seabed authority, an issue that looked ripe for further development at the next session. The problem of how to compensate the landlocked and geographically disadvantaged states appeared to be settled by the enormous majority in favor of fisheries sharing and the adamant refusal of coastal states to share continental shelf resources. The landlocked and shelflocked states confronted a rather stark choice on the offer of trading off their shelf demands in return for fisheries sharing. As a group they were too dependent on the conference format (they had no unilateral option), and therefore were too weak to fight this compromise to the bitter end; their acceptance of it appeared to be only a matter of time. Adherence to the shelf resources sharing demand might at most still help them to wring a concession from the margin states on revenue sharing with the seabed authority from exploitation of the margin beyond 200 miles.

Nevertheless, although Caracas saw substantial progress toward agreement on the continental shelf, this progress was rendered vulnerable by the lack of agreement on other issues with which the continental shelf was connected in the coastal zone package. No agreement was in sight on key issues like coastal state control over fishing, scientific research, and pollution control in the zone. The question of passage through straits also remained extremely controversial. Failure to agree on any of these issues threatened to ruin the package deal of which the continental shelf negotiations were only a part.[11]

International Seabed Regime and
Machinery Questions

In C.I, the debate on the international seabed regime and machinery moved away from the stalemate over a licensing versus an operating regime and machinery (the question of "who may exploit the area," Article 9 of the draft articles before the committee) and toward the closely associated but more specific matter of the rules and conditions under which exploration and exploitation activity could occur. It was quite generally felt in the committee that these twin questions constituted the nub of the problem facing the delegates, and that if answers could be devised for them, then all other aspects of the regime and machinery would fall smoothly into place. The transition to this more specific level of debate in C.I was substantially facilitated by the progress on limits in C.II. The narrowing of the limits question to a 200-mile/margin framework enabled C.I to bypass one of the quandaries that had paralyzed the negotiations for so long—how to discuss the regime before the area, and therefore the resource content, was known. The higher priority given to coastal zone matters because, among other things, of the oil crisis, meant that the dilemma of whether to define the boundary or regime first was decided in favor of defining the boundary first for reasons largely external to the international regime and machinery question.

The general debates during the first few weeks of the Caracas session merely repeated the operating versus licensing positions already clearly established in the SBC, and contributed little to C.I's work. Later in the session, however, the discussion moved away from the well-established generalities of licensing versus operating, and concentrated on more specific aspects of Article 9: "Who shall exploit the area?" Four variants existed for this article in the alternative texts from SC.I of the SBC, but these merely enshrined the different points of view (licensing, operating, mixed) and did not offer an acceptable path around the impasse. Not until late in July did the work of C.I break out of the SBC's mold by concentrating on compromise formulations for Article 9.

The Group of 77 started the movement by beginning work on a compromise draft representing the position of the Group as a whole, but Jamaica was first in the field with a proposal.[12] The Jamaican proposal sought to open a new path by avoiding the question of who shall exploit and concentrating instead on developing guidelines for the regulations that would govern exploitation. The guidelines were presented as a set of principles designed to protect the various interests of the opposing sides in seabed exploitation, including control by the authority, wide and equitable participation, maximum benefits for humanity, and security of resources supply. The approach was novel, and had the advantage of offering an alternative to "who shall exploit," but it did not resolve the basic conflicts so much as offer to refer them to the authority. From SBC experience this tactic would have been likely simply to revive the whole debate over whether the council or the assembly should dominate the authority. In any event, the Jamaican proposal was quickly relegated to the sidelines despite the fact that it had the support of the chairman of the informal meetings and had received a favorable response from the United States as a basis for negotiation.

The reason for its relegation was that the Group of 77 succeeded in presenting a unified proposal of its own which immediately dominated the proceedings.[13] Jamaica insisted that its proposal was complementary to that of the 77, but it failed to gain a sympathetic hearing from the Group and its proposal was not pursued. The Group of 77 proposal stuck with the "who shall exploit" approach, but attempted to open up grounds for compromise with the states possessing nodule mining technology. Its text read:

> All activities of exploration of the Area and of the exploitation of its resources and all other related activities including those of scientific research shall be conducted directly by the Authority.
>
> The Authority may, if it considers it appropriate, and within the limits it may determine, confer certain tasks to juridical or natural persons, through service contracts, or association or through any other such means it may determine which ensure its direct and effective control at all times over such activities.

While paragraph one reflected the developing countries' strong preoccupation with ensuring the paramount rights and powers of the authority, paragraph two left sufficient room for meaningful negotiations with the developed states. Agreement on this draft marked a big step forward for the Group of 77 in C.I, and gave it the initiative in the proceedings. Since the draft had the support of more than two-thirds of the states at the conference, the most obvious next step for C.I was to explore the leeway within it in search of an acceptable compromise with the developed states. Surprisingly, the United States accepted the Group of 77 text as a basis for negotiation, and U.S. draft of Article 9, presented shortly after the 77's draft, was largely oriented toward eliminating the direct reference to types of contract that the United States found objectionable in the 77's text.[14] The Soviet Union, Japan, and the European states were generally unenthusiastic about the Group of 77's version of Article 9.

Ironically, the very strength of the Group of 77's proposal provided the developed states, particularly the United States, with the means to regain the initiative in C.I. The massive support behind the 77's Article 9 meant that the 77's text was really the only realistic basis on which negotiations on this crucial point could proceed. The developed states implicitly accepted this, and began to deluge the Group of 77 with questions and requests for clarification concerning the text. Many of these inquiries were technical in nature, and reflected the concern of the developed states with the specific rules and conditions of exploitation. The Group of 77 had great difficulty coping with this tactic for two reasons. First, the Group contained no delegations with the technical expertise necessary to handle a debate on that level. Second, the text represented a delicate compromise within the Group of 77 itself, and this made it almost impossible for the Group to bargain over the text without destroying its internal consensus. The Group could thus do little more than hold its ground and insist that the text as given was capable of sufficiently flexible interpretation to meet the needs of all delegations.

By the end of the first week in August, this confrontation looked like becoming the new point around which the C.I negotiations would stall. The chairman of the informal meetings made a largely unsuccessful attempt to get the debate onto more creative ground by introducing on August 7 a list of subjects for discussion. But on the next day the U.S. delegate seized the initiative in a long and strongly worded speech that was one of the dramatic highlights of the whole session. He demanded that the committee deal with the conditions of exploitation and rejected what he interpreted as the efforts of the Group of 77 to avoid that topic by refusing to enter into more detailed discussions of Article 9. He presented a long list of topics that his delegation insisted must be discussed, and on many of them made it clear that the U.S. position was not negotiable. His speech was heavily loaded with the rhetoric of free enterprise, and he did not take pains to conceal his country's determination to reject any seabed regime and authority that failed to define clearly the conditions of exploitation.

In diplomatic terms this U.S. intervention was exceedingly heavy-handed, and it provoked acrimonious and time-consuming ideological exchanges between developed and developing countries. It seems likely that the tone of the U.S. delegate's speech related at least as much to internal bureaucratic politics of U.S. law of the sea policy-making as it did to the real ideological differences within C.I. Despite these unhelpful aspects, however, the U.S. initiative succeeded in breaking the deadlock on Article 9 and turning the discussion toward the conditions of exploitation. In so doing, it opened the way for discussion of the essential points of difference in C.I, and thereby provided some prospect that the Group of 77's Article 9 could serve as the basis for a compromise acceptable to all.

The United States followed up its initiative by introducing a formal document containing its proposals for conditions of exploitation. This in turn stimulated proposals on the same subject from eight of the EEC states, Japan, and the Group of 77.[15] The Group of 77 took nearly a week to negotiate its conditions of exploitation draft internally, during which time the work of C.I came to a halt. The painful production of a draft by the Group indicated its acceptance of the U.S. ultimatum and demonstrated once again the powerful position that possession of technological capability gave to the Americans. In the event, the Group of 77's text gained support from a number of states outside its ranks, including China, Albania, Norway, Sweden, Spain, and Romania.

The three proposals from developed countries reflected essentially the same approach to the problem. Their differences were relatively minor ones of emphasis, detail, and style, and for the purposes of comparison with the Group of 77's proposal they can be treated as a group. The Soviet Union did not make a proposal on conditions of exploitation at Caracas, but its concern with the right of states themselves to exploit seabed resources aligned it quite clearly with the other advanced technology countries, particularly the EEC group.[16]

The approaches taken by the two groups of states to conditions of exploitation were based on wholly different perceptions of the problem. The advanced technology countries, stressing maximum accessibility to seabed resources, concentrated their attention on defining the rights and duties of the entities engaged in exploitation (whether states or companies) and on the way in which the activities of those entities could be supervised by the authority. Their approach was to define conditions, the fulfillment of which would automatically result in the allocation of specified rights and duties to entities desiring to engage in exploitation. Such a system would place nearly all powers of initiative in the hands of the exploiting entities (the advanced technology states and their companies) and leave the seabed authority a largely passive supervisory body.

The developing countries, by contrast, followed the lead of their draft on Article 9 and put great emphasis on defining the rights and powers of the authority. They wanted the authority to retain virtually all powers of initiative, and not to be hamstrung by detailed, inviolable conditions within the convention. In particular, they wanted to avoid getting bogged down in detailed technical provisions beyond their competence, and so tried to keep the discussion of conditions of exploitation on the more general level of guiding principles. They also gave high priority to preventing detrimental effects on the earnings of developing country mineral exporters from competition with seabed exploitation. This required them to give much less weight to maximum accessibility than did the advanced technology states, and also to advocate power for the authority to regulate seabed mineral production.

The enormous perceptual gulf separating the two groups in C.I obviously persisted despite the shift to discussion of conditions of exploitation. The shift was nonetheless an important landmark in the committee's progress because it moved the debate to a level of specificity on which the delegates could define and clarify their points of difference in a context that offered reasonable chances for solutions. In effect, discussion of the conditions of exploitation broke up the central question "who shall exploit?" into a whole package of smaller and more manageable sub-questions. Comparison of the four proposals revealed agreement only on a few relatively minor items such as the applicability of rules to exploitation, the power of the authority to suspend operators found violating the convention, and the necessity for operators to have some guarantees of security of tenure. Indeed, perhaps the most important agreement revealed in these texts was the implicit strengthening of the Group of 77's reluctant concession that if any exploitation of the common heritage were to take place, it would have to be undertaken at least for the first several years by the private companies of the Western states that possessed the necessary technology and expertise.

Major disagreements in the texts of the two groups were legion, the most important being as follows.[17]

(1) The 77 wanted the title to the area and its resources to be vested in the authority, while the Americans opposed this as unnecessary, and wanted to give the authority simply rights to supervise exploitation activity.

(2) The 77 wanted the authority to control scientific research, processing, and marketing functions, as well as exploration and exploitation. The developed states wanted the authority's jurisdiction limited to exploration and exploitation.

(3) The 77 wanted discretionary powers for the authority in choosing operators and in allocating both mine sites and contracts for various stages of activity. The developed states wanted a virtually free right to explore and prospect, and automatic allocation of exploitation rights to those who fulfilled set conditions. In addition, they wanted operators to have the right to choose mine sites within the limits set by the convention.

(4) The 77 wanted discretionary powers for the authority in any transfer of rights from one operator to another. The developed states

wanted an essentially free right of transfer dependent only on fulfillment of specified conditions.

(5) The 77 wanted the authority to have financial and administrative control of all operations without assuming any of the financial risks. The developed states rejected this, and wanted the financial responsibilities of operators limited to specified schedules of work (to avoid speculation in mine sites) and to payments to the authority on the basis of formulas related to activity and production.

(6) The 77 wanted the authority to have the right to change the terms of a contract, including suspension or termination, in the event of "a radical change in circumstances". This was wholly unacceptable to the developed states on the basis of their arguments for security of tenure and investment.

(7) The 77 wanted the authority to have the power to impose production controls, a position that was unacceptable to the United States.

(8) The 77 appeared to see the authority as a sovereign body, whose decisions, once taken, would be irreversible except by itself. The developed states, particularly the United States, wanted some form of tribunal to handle disputes between the authority and operators.

(9) The 77 wanted extensive information transfers from operators to authority as well as commitments on training programs for, and employment of, personnel from the developing countries. The developed states were much more cautious about the release of proprietary information, and, except for Japan, made no suggestions on training and employment.

The four texts on conditions of exploitation thus made clear the continuing fear among the Group of 77 that the developed states would use their technology to loot the common heritage, and the corresponding fear among the advanced technology states that an authority controlled by the Group of 77 would deny them access to secure supplies of mineral resources from the seabed. The texts also emphasized the conflict between the developing countries' great dislike for technical details, and the developed states' insistence that firm understanding could only be had by specifying the technical aspects of agreements.

Detailed negotiation on all these points was only beginning by the end of the Caracas session. In the last week of August C.I set up a special working group on Article 9 and the conditions of exploitation, but this had time for only a handful of meetings, and suffered from the fact that it was too close to the end of the session to begin breaking new ground. As a result of the Caracas session, then, the work of C.I appeared to be leaving the declaration, definition, and clarification stage and entering the negotiation and compromise stage. As the chairman of the working group phrased it, "The Committee had established the opposites of the dialectic and was about to commence the task of unifying and reconciling them."[18]

Despite the lack of negotiation in C.I at Caracas, the committee made important advances in relation to what had been accomplished in SC.I of the SBC. In terms of the seven factors identified in Chapter 7 as being responsible for the slowness of the committee's progress, however, the situation after Caracas was not much improved. Only one factor was effectively removed as a barrier to the committee's work at Caracas, and that was the uncertainty over the definition of the area. All the other six factors either remained in force as before, or else became even more formidable obstructions to progress.

First, the innate slowness of the consensus procedure was carried over from the SBC to the conference as a result of the debate on rules of procedure. This largely nullified the higher pressure to reach agreement normally associated with the conference format, and was exacerbated by the 50 percent increase in the number of participating states.

Second, the polarization of alignments between developed and developing countries became even more acute as a result of the Group of 77 reaching agreement on common positions. The ideological tones of the debate on conditions of exploitation aggravated this condition, as did the move toward a common position of the major industrialized states.

Third, incentive-taking by states continued to be a live option in the background, in particular with relation to the Metcalf bill in the United States, whose proponents ensured that the bill remained before Congress.

Fourth, economic and technological developments in the nodule mining industry continued to raise tensions as the industry moved inexorably toward an operational capability. The Secretary-General submitted to the session a 92-page update to his previous reports on the subject.[19] This contained the expected news of steady progress in the technology of mining and processing, and also attempted to outline information and options on contentious issues such as the likely impact of nodule mining, the determination of revenues for the international machinery, and the question of how to prevent adverse effects from nodule exploitation on existing producers.

The report also contained evidence of another kind of development which, along with progress in the technology, appeared likely to add to the difficulties of C.I's work. This was the news that many of the companies interested in nodule mining were beginning to pool their resources in large multinational consortiums. Two of these were announced. One involved Tenneco, one European, and three Japanese companies; the other involved Kennecott, two British, one Canadian, and one Japanese company. Another was expected to follow involving INCO, several German, one American, and possibly several Japanese companies.[20] This movement was partly explained by the stage the nodule mining industry had reached. Large amounts of money were needed to finance the research and development necessary to bring the technology to an operational level, and forming consortiums like these helped to spread the load. The tendency for these consortiums to be centered on, or at least to include, a U.S. company, however, carried possibly ominous political overtones. While the U.S. companies were acknowledged leaders in the technology, they also worked under the flag of the one country that had shown willingness to consider unilateral legislation to protect nodule miners. Consequently, this pattern of alignment among the mining companies offered every prospect of appearing threatening to the developing countries, and thus raising further the tensions within the negotiations from this source.

Fifth, and perhaps most important as an obstacle to progress, was the situation by the end of the Caracas session relating to problems arising from the genuine complexities of the issues before C.I. Some of these problems were ongoing from the SBC, but three were additions, consisting either of problems that had been realized previously, but which became prominent at Caracas, or else of new problems arising out of the Caracas discussions.

Among the ongoing problems, two were of notable importance at Caracas: the influence of commodity matters, already discussed above, and the politicization of information relevant to C.I's work. The politicization of information continued to have its most disruptive effect in relation to the linked questions of the likely impact of nodule mining and the possibility of adverse effects on existing developing country exporters. At Caracas, debate on these questions was largely concerned with the issue of whether or not the authority should be given powers to control the volume or pricing of seabed mineral production. The developing countries had long favored such powers, while the United States opposed them vehemently, and at Caracas studies were presented that offered opposing interpretations on these questions in line with the policy preferences of the two sides. A series of UNCTAD papers supported the developing countries' argument that the danger of nodule mining disrupting existing metal markets was sufficient to justify the exercise of stringent production controls by the authority.[21] The developing countries also used the Secretary-General's report to back up their contention that high revenues would accrue from seabed mining. Again the danger arose that in its most extreme form their position looked like opposition to any deep seabed exploitation, an appearance that was enhanced by frequent assertions that nodule mining would benefit primarily the resource-importing developed states.

The advanced technology states spent much time refuting the arguments in the Secretariat and UNCTAD documents and trying to make the case that increased sources of supply lowered prices and therefore benefited developing country importers as well. The United States eventually submitted a document of its own arguing that the impact on existing producers would be minimal and that production controls were not under any circumstances a practicable approach to the problem.[22] This ritual continued to be based on highly speculative data, but the sustained association of information with what amounted to ideological positions strengthened the tendency for all information to be politically suspect, and thereby encouraged states to cling to their predispositions on the matter. This process promised to create endless difficulties when the negotiations reached a sufficiently detailed level, as they appeared to be doing on the conditions of exploitation, to require decisions based on the best available information.

The three additional problems all arose from other issues in the law of the sea negotiations: high seas fishing, the regime of islands, and the transfer of technology. The high seas fishing problem was yet another reincarnation of the temptation to link seabed jurisdiction with jurisdiction in the superjacent waters. Fourteen developing countries, the majority of them speaking in C.I, declared themselves at Caracas in favor of various kinds of increased international control over high seas fishing, and some of them suggested that this power be vested in the international seabed authority. A move in this direction was also desired by the 41 members of the OAU, which had favored the establishment of an international sea fisheries regime in their 1973 and 1974 declarations. Strenuous opposition to this idea came from seven states, all with fishing interests, including the United States and the Soviet Union, both of which adamantly rejected any extension of international authority to the water column. By the end of the session it was abundantly clear that any decision by the developing countries to push this matter would evoke extreme hostility from the two superpowers. Both were primarily concerned in this issue with assuring complete freedom for their navies against any interference from the authority under the "peaceful uses" part of the convention. Only secondarily was their opposition related to fishing interests, although this could not be discounted, especially in the case of the Soviet Union.[23]

The problem of the regime of islands emerged largely in the context of C.II. Within the framework of the debate on coastal state jurisdiction it became increasingly clear at Caracas that the application of a 200-mile zone to every rock and reef in the ocean would result in enormous areas of seabed and superjacent water falling to island and island-owning states. This had always been clear in theory, but progress on the economic zone at Caracas finally began to bring the problem into prominence within the negotiations. Disputes arose over whether archipelagos belonging to coastal states should have the same rights as oceanic archipelagos, and more important, over whether all islands should be given full maritime zones, or whether and how restrictions should be applied to rocks, reefs, and uninhabited islands. Division was mostly along geographical lines, except for the hotly contested issue of the large number of strategically placed islands under colonial control, where divisions were predictably ideological.[24]

For C.I, the problem was that if the island and island-owning states, including the colonial powers, should win their case, either within the negotiations or by taking unilateral action, then some of the seabed areas within their control, particularly in the Pacific, would contain ore quality manganese nodules. This would deprive the seabed authority of monopoly control over nodule resources, and require that it compete with other resource holders to attract investment from entities capable of deep sea mining. Such a development would wreck the type of authority envisaged by the Group of 77 and consequently could shatter the carefully constructed basis for negotiation in C.I.

The problem of transfer of technology has been alluded to in the discussion on disputes over conditions of exploitation in C.I, but in fact the main activity on this issue took place in C.III. The problem centered on a demand made by 18 states representing a cross section of the Group of 77 that "all blueprints and patents of the equipment, machinery, devices and processes used in the exploration of the international area, the exploitation of its resources and related activities be made available to all developing States upon request."[25] This proposal reflected a growing feeling among the developing countries that with the decline in projected revenues from the international seabed area, the acquisition of technology was one of the most important benefits likely to accrue to them from the common heritage. The problem was that if the exploiting entities were to be mostly Western mining companies, as seemed inevitable, then they would be exceedingly reluctant to make proprietary information of that kind public. They would view a requirement along such lines as a breach of their investment security, and could very probably get the support of their governments in opposing it. The danger was that a decision by the developing countries to push this issue in C.III would have a disruptive effect on the integrated package under negotiation in C.I.

Sixth, and finally, among the factors continuing to impede the work of C.I was the fact that the attention of the conference remained rooted in C.II on the question of the economic zone, leaving C.I as something of a sideshow in terms of states' priorities.

Given the advances made in C.I at Caracas, the question at the end of the session was whether or not this progress was sufficient to offset all the numerous factors still tending to slow down the committee's work. No definitive answer to this question could be given, but the grounds for optimism about the committee's work were not large. The failure of the session to reach any substantive conclusions allowed the advanced technology states another seven months during which to strengthen their options. It remained to be seen whether the Geneva session would make the necessary breakthrough on the international seabed regime and machinery negotiations, or whether it would merely find yet another formula for expressing the fundamentally opposed positions of the two sides.

The Groups

Changes in the composition, strength, and level of activity of the negotiating groups at Caracas were almost as important as shifts on policy alignments.

The Group of 77 emerged from Caracas as a stronger actor than it had been in the SBC, but it still displayed great weakness and disarray in the crucial forum of C.II. In C.I, the Group of 77 overcame its internal differences with the two documents on Article 9 and conditions of exploitation put forward by the Group as a whole. Its unity in this forum was considerably bolstered by the commodity producer and new economic order movements, which not only intensified the Group's symbolic role for its members, but also provided radical leaders like Algeria and Tanzania that were determined to build the Group's unity around large ideological issues. In C.III, the Group made substantial progress toward agreement on scientific research (62/C.3/L.13), but was plagued here by splits between its coastal and noncoastal members. In C.II, however, the Group made no progress toward common positions on any key items. As in C.III, the primary split was between coastal and noncoastal members, but there were also splits based on other geographical or regional criteria on issues such as islands, straits, and the margin. For a time it looked as if the Group of 77 might be uniting around the issue of emplacement of military devices on the continental shelf.[26] But sharp warnings from the United States that it was not prepared to discuss this subject at the conference dampened down what threatened to become a major confrontation between the developing countries and the great powers, and the former decided not to risk the conference by pressing the matter.[27] Despite its internal differences, which were enhanced by the influx of developing countries into the negotiations, the Group of 77 retained its strong symbolic importance for the developing countries as a focus of activity at Caracas.

The Latin American group and subgroups showed a much lower profile at Caracas than they had done in the SBC. Only one distinctively Latin American proposal was made at Caracas,[28] and for most Latin American states the Group of 77 provided the primary focus of sponsoring activity. In addition, there was an increase in the number of states sponsoring proposals singly or in combination with ad hoc groups that crossed regional or political lines. The two South American landlocked states concentrated their activity almost entirely within the landlocked and geographically disadvantaged state group.

By contrast, both the African and Arab groups showed a somewhat stronger presence at Caracas than they had done in the SBC, although the Group of 77 continued to be the main focus of sponsoring activity for most of the members of both groups. Two proposals came out of the African group at Caracas, and one, the first for the group, from the Arab group.[29] Furthermore, there was a notable increase in the number of states in both groups sponsoring proposals alone or with ad hoc issue-specific groups. The African group maintained the strong regional position it had established in its declarations but still showed signs of internal divisions on a number of issues, particularly those between coastal and landlocked states. The Arab group, helped by the influx of new states into the conference, emerged as a separate entity for the first time, distinct from both the African group and the Group of 77.

The behavior of the Asian states remained unchanged from the SBC to Caracas. No Asian group came into being, and the Asian states continued to give their primary loyalty to the Group of 77, while also remaining active as single state and issue-specific group sponsors.

Among the developed states, the Soviet group continued to act as an important focus for its members, and the EEC group emerged as an active player for the first time. The Soviet group showed a stronger united front at Caracas than it had done in the SBC. Nearly all the sponsoring activity of its members occurred within the context of the group's six proposals,[30] with only Romania maintaining an independent stance. The EEC group produced three proposals, two with eight of its members and one with all nine,[31] but most of its members also kept up active single and ad hoc group sponsoring.

The Western states as a whole continued to concentrate their activity on single and issue-specific group sponsoring, although they were not as dominant in this sphere as they had been in the SBC. The United States delegation retained its character as a miniature model of U.S. internal bureaucratic politics, and although it still concentrated on single sponsored proposals, did participate in two issue-specific cosponsorings. The Americans appeared to conduct most of their negotiations through bilateral contacts.

Issue-specific groups in general suffered a drop in strength compared with the later years of the SBC. Several, including a shipping states group and a maritime powers group, remained active but produced no proposals. The two relevant to seabed politics were more visibly active, but of these only the landlocked and geographically disadvantaged state group gained in strength. Short-lived, ad hoc issue-specific groups flourished in considerable numbers at Caracas, but very few produced cosponsored proposals.

The landlocked and geographically disadvantaged group showed much greater coherence at Caracas than in the SBC. It produced three proposals sponsored by many more states than had been its best efforts in the SBC,[32] and most of its 28 visibly active members made the group the exclusive focus of their cosponsoring activity in Committees II and III. Some signs of regional divisions within the group cropped up in the form of proposals sponsored by the landlocked states of a particular continent,[33] but in general this group was more prominent for splitting, rather than being split by, regional or political alignments. It even extended its activity to C.I by putting forward a consensus draft accepted by 50 states,[34] though this was on a minor matter and did not threaten the dominance of the Group of 77 in that committee. The increase in strength of the landlocked and geographically disadvantaged group at Caracas could not, however, hide the fundamental weakness of the group's power base and its total dependence on the international negotiations for influence. The main drift of the session was not in the group's favor, and its unity therefore had overtones of desperation.

The coastal state group, by contrast, scarcely survived the Caracas session as a coherent entity, and virtually ceased to be an active player at the conference. Coastal state interests as a whole fared well at the session, and individual coastal states remained powerful actors, but the group itself was unable to reach a unified position, and because of this appeared to reach the limits of its utility as a vehicle for the policies of its members. The major proposal put forward by nine of the group's members received an unenthusiastic, and at times hostile, reception in the plenary meeting, and this event appeared to break the group's momentum.[35] Its members shifted the focus of their activity elsewhere, such as to ad hoc groups[36] and to more broadly based groups the purpose of which was to work out compromise positions among contending alignments.

The decline of the coastal state group as a group, however, accompanied major moves toward realization of many of its policy aspirations at the session, particularly those relating to the economic zone, the continental margin, and archipelagos. This fact supports the view that the group did not so much collapse because of its internal divisions as decline in a natural, evolutionary way on the realization that its prime purpose had been fulfilled. Once the coastal state position had been firmly established, as it was by the end of the Caracas session, then the need was not for a group to continue advocating it, but for more conciliatory efforts aimed at making the position acceptable to noncoastal states.

This interpretation, while valid in many respects, does not hide the fact that the coastal states were seriously divided among themselves on many aspects of policy. Differing schools existed on the economic zone as well as on the continental margin, and the absence of a moderate coastal state group strengthened the centrifugal forces working against a unified coastal state position. One of the more visible illustrations of this was the strengthening of moves toward a territorialist group among the 17 states identifiable as members of a territorialist school at Caracas.

Perhaps the most important development among the negotiating groups at Caracas was the rise to prominence of what might be called "conference interest groups." As the name implies, these groups were not built around one policy interest in competition with others, but were created with the aim of finding compromise positions on key issues before the conference. Two of these assumed central roles at Caracas, one dealing with the specialized but crucial issue of settlement of disputes, and the other, convened by the Norwegian Jens Evensen, dealing with the major questions before C.II and C.III.

The disputes settlement group, formed during the Caracas session, contained about 30 members including the United States and both developed and developing, and landlocked and coastal, states. At the end of the session it produced 62/L.7, a draft proposal on settlement of disputes.

The Evensen group was formed during 1972 at the SBC, and at Caracas was attended by representatives from a maximum of 26 states. It contained all the major maritime powers and most of the major coastal states, both developed and developing, as well as many of the leaders, and all the developed states, from the coastal state group. Its purpose was to try to initiate the bargaining and compromising process, but given the stage of the negotiations at Caracas, the group did not find itself in a position to make a significant move. Although it did become more prominent at the session after the coastal group failure with 62/L.4, the group was clearly aiming at the next session to make its play. The major problems confronting the Evensen group, aside from the intrinsic difficulty of finding compromises, were the almost complete absence of landlocked and geographically disadvantaged members, and the suspicion with which the group was viewed by many members of the Group of 77 on the grounds of its elite membership and approach. The emergence of these conference interest groups at Caracas was an important sign that cooperation was developing as a parallel to confrontation as a method of work within the negotiations.

By the end of the Caracas session, then, the Law of the Sea Conference had reached an important psychological turning point. According to Resolution 3067, it was supposed to finish its work by the end of 1975, a deadline on which the United States in particular had placed considerable emphasis.[37] Because the Caracas session had not made the substantive progress originally expected, this meant that all attention was turned to the Geneva session as the place at which the necessary breakthroughs on substantive matters would occur. Most of the deficiencies in the SBC's preparatory work had been rectified at Caracas, and therefore there was no longer any procedural barrier to negotiations. Furthermore, the clarification of issues and positions in Committees I and II, and the emergence of the Evensen group as a forum for compromise, all appeared to help set the stage for fruitful bargaining.[38]

Much ado was made of the danger posed to the negotiations by the possibility of states taking unilateral action during the intersession period. While this danger was real enough, particularly in the case of key states like the United States, Canada, Norway, and Britain, the main threat to the negotiations came not from without but from within. The inability of the conference to overcome the slowness of its own process was demonstrated graphically by the whole Caracas session, and unless events took a radically different turn at Geneva, the conference looked very likely to overrun its own deadline. Should that happen, there seemed every likelihood that it would lose the drive and urgency imparted by a strict working deadline and lapse back into the interminable deliberations of SBC days.

NOTES

1. On the subject generally, see C. F. Bergsten, "Commodity Shortages and the Ocean," in National Science Foundation, Perspec-

tives on Ocean Policy (Washington: Johns Hopkins University Press, 1974), pp. 167-78.

2. A/CONF.62/23.

3. See Ch. 8, n. 28.

4. A/CONF.62/30/Rev.1, p. 18.

5. A/CONF.62/SR.1-13, pp. 44-50; and A/CONF.62/6 and 7.

6. A/CONF.62/SR.1-13, p. 9.

7. Iraq, Sweden, Poland, Hungary, and the Byelorussian republic did not renounce their SBC positions; Afghanistan, Belgium, Kuwait, the Netherlands, and Nepal reaffirmed their opposition to the 200-mile and margin alternatives; and Bhutan and West Germany joined the school as newcomers to the negotiations. It is significant that a majority of these states were developed and that no African or Latin American states were included.

8. The eight were the United States, Denmark, Jamaica, Trinidad and Tobago, Ghana, Ethiopia, Mauritius, and the Malagasy Republic.

9. A/CONF.62/C.2/L.39.

10. A/CONF.62/C.2/L.47.

11. See, for example F. X. Njenga in Perspectives on Ocean Policy, op. cit., pp. 87-105, who argued that the African states saw the economic zone as a national zone in which all rights, except those defined as belonging to the international community, would reside with the coastal state. This was a far cry from the efforts of many developed states to define the zone the other way around.

12. C.1/CRP.3, and in a truncated version in A/CONF.62/C.1/L.3, p. 19.

13. C.1/CRP.4, and ibid., alternative B of Article 9, p. 6.

14. C.1/CRP.5, and ibid., alternative C, p. 7.

15. United States, A/CONF.62/C.1/L.6; EEC group minus Ireland, A/CONF.62/C.1/L.8; Japan, A/CONF.62/C.1/L.9; and the Group of 77, C.1/CRP.7 and A/CONF.62/C.1/L.7. On the negotiations up to this point, see also A. O. Adede, "The System for Exploitation of the 'Common Heritage of Mankind' at the Caracas Conference," American Journal of International Law 69, no. 1 (January 1975): 31-49.

16. A/CONF.62/C.1/SR.8, pp. 7-10, and interviews with delegates. See also M. I. Lazarev (trans. T. Sulikowski), "Scientific-Technological Progress and the Search for Legal Regulation of Possible Seabed Uses," Ocean Development and International Law 3, no. 1 (1975): 69-86.

17. A comparative table of the four documents was prepared for the Geneva session as C.1/CP/Working Paper no. 2, March 18, 1975.

18. A/CONF.62/C.1/SR.14, p. 6.

19. A/CONF.62/25.

20. Ibid., p. 17, and interviews with delegates.

21. TD/B/449, 449 Add.1, 483, and 484; and TD/113/Supp.4. These were summarized in A/CONF.62/C.1/L.2. Chile also submitted a paper that reflected the Group of 77's line of argument: A/CONF.62/C.1/L.11 and Corr.1.

22. A/CONF.62/C.1/L.5.

23. A/CONF.62/C.1/SR.6, p. 2; SR.7, p. 6; SR.8, p. 8; and interviews with delegates.

24. A/CONF.62/C.2/SR.36-40. For a fascinating study of this problem in relation to the Pacific Ocean, see F. M. Auburn, "Some Legal Problems of the Commercial Exploitation of Manganese Nodules in the Pacific Ocean," Ocean Development and International Law Journal 1, no. 2 (Summer 1973): 185-200.

25. A/CONF.62/C.3/L.12. An earlier proposal by Nigeria (A/CONF.62/C.3/L.8) expressed similar demands.

26. This issue was first raised by Mexico, and then by Mexico and Kenya together, in A/CONF.62/C.2/L.42 and Rev.1. By the end of the session 34 states had joined them in sponsoring a proposal that "No state shall be entitled to construct, maintain, deploy or operate on or over the continental shelf of another state any military installations or devices or any other installations for whatever purposes without the consent of the coastal state." This wording clearly applied to submarines, and so was exceedingly contentious. Mexico's interest in this issue went back to negotiations in the ENDC and the CCD, and related to its claim to sovereignty over the shelf. See E. D. Brown, Arms Control in Hydrospace: Legal Aspects (Washington, D.C.: Woodrow Wilson International Center for Scholars, 1971), p. 99.

27. Interviews with delegates.

28. A/CONF.62/C.2/L.58, sponsored by 14 Latin American states on the question of islands.

29. A/CONF.62/C.2/L.44, L.62 and L.82.

30. A/CONF.62/C.2/L.11, L.26, L.27, L.38, and L.52; and C.3/CRP./Sci.Res./15.

31. A/CONF.62/C.1/L.8, A/CONF.62/C.2/L.40 and L.54.

32. A/CONF.62/C.2/L.33 and L.39, and A/CONF.62/C.3/L.19, sponsored respectively by 22, 22, and 16 states from all regions.

33. A/CONF.62/C.2/L.65 and L.76 by Bolivia and Paraguay, and A/CONF.62/C.2/L.45 Rev.1 by Botswana, Lesotho, Uganda, and Upper Volta.

34. A/CONF.62/C.1/L.3 Corr.6, and A/CONF.62/C.1/SR.17, intervention by Bolivia.

35. A/CONF.62/L.4, sponsored by Canada, Chile, Iceland, India, Indonesia, Mauritius, Mexico, New Zealand, and Norway.

36. A/CONF.62/C.2/L.66, and A/CONF.62/C.3/L.6.

37. See, for example, J. R. Stevenson and B. Oxman in American Journal of International Law 69, no. 1 (January 1975): 30.

38. For a detailed account of the New York and Caracas negotiations, and also of some negotiating problems, see Edward Miles in Francis T. Christy, et al. (eds.), Law of the Sea: Caracas and Beyond (Cambridge: Ballinger, 1975), pp. 37-94.

CHAPTER

10

THE THIRD UN CONFERENCE ON THE LAW OF THE SEA 1974-75, PART II: GENEVA

Government reactions to the Caracas session were fairly uniform. In general they indicated a combination of disappointment in terms of initial expectations and satisfaction with what was actually accomplished. There was widespread affirmation that the session had not been a failure and that governments were prepared to continue supporting the conference as the best road to a new international law of the sea.

Two factors confirmed this continuation of support: the sustained activity in bilateral talks and among the informal negotiating and consultative groups during the intersession period, and the restraint on the matter of unilateral action from those governments under pressure to increase their maritime jurisdiction. One promising sign was the fact that many of the informal groups meeting to discuss the law of the sea during the intersession period were conference interest groups or other groups crossing the main lines of division in the negotiations. These included the Evensen group, the disputes settlement group, a group on marine pollution, a group on C.I matters, and a meeting of the Commonwealth states. No tangible results came from these meetings, but most participants felt that the exchange of views and clarification of positions were useful. Some of the coastal state group members tried to form a group on fisheries questions, but this appears to have been absorbed into the Evensen group. The OAU met in February 1975, but added nothing to its existing declaration, and indeed experienced a weakening of its internal compromise on the issue of landlocked states participating in the economic zones of their coastal neighbors. As revealed at Caracas, at least six African landlocked states—sponsors of C.2/L.39—were not content to accept the fisheries sharing offer but wanted rights to continental shelf resources as well.

The only concrete results from an intersession meeting came from the Group of 77's conference on raw materials in Dakar during February. This did not discuss the law of the sea directly, but it did add weight to the commodities movement, which was playing an increasingly important role in seabed politics. Steps were taken in the direction of linking the new economic strength of the oil-exporting states to the price-fixing efforts of other developing commodity exporters. These steps bolstered the unity of the Group of 77 on commodities issues and the new economic order, and in so doing seemed very likely also to increase its unity of purpose and its interest in C.I.

Restraint on unilateral action during the intersession period was considered crucial by most observers of the conference. Without such restraint, especially from major states, the real, but not formally agreed upon, progress made at Caracas would be in grave danger, and the conference itself might well be overtaken by a wave of unilateral claims made in response. Restraint by the United States was especially vital in this regard, and also especially precarious because of the peculiar separation of powers in American government.

In the event, although restraint was not universal, no major breaches occurred. The successful liberation movement in Guinea-Bissau added its name to the list of African territorialists with a claim of 150 miles, while Oman claimed a pollution control zone of 50 miles, but both of these were minor states and their actions had little impact. A more serious threat came from Norway, which made moves toward a 50-mile fishing zone scarcely a month after the end of the Caracas session. Active Norwegian diplomacy with the affected states helped to mitigate the impact of the claim, but given the central role of Norway at the conference it was not a good omen for the international negotiations.

In the United States, attention focused on the Fisheries Extension Bill (S.1988), a proposal arising, like the Hard Minerals Bill, from within the Congress without the approval of the executive. S.1988 was aimed at extending U.S. fisheries jurisdiction out to 200 miles, and only after much lobbying was the executive able to defer the bill by arguing its disruptive effect on the international negotiations. This line of argument, like that successfully used against the Hard Minerals Bill, left the executive increasingly dependent on a rapid and favorable conclusion of the Law of the Sea Conference. Another development, possibly more threatening, was the posting of a notice by Deepsea Ventures of a claim to a nodule mine site in the Pacific. The company requested United States diplomatic protection and also protection for its investment, and sent notices of its claim to over a dozen foreign capitals. Although there was surprisingly little reaction to this claim, it represented a clear statement from the nodule mining industry that their technical and organizational capabilities were developing apace and that they would not be prepared to wait on the conference for an indefinite length of time.[1]

The mere passage of time, as it had done for more than the preceding decade, allowed local situations to develop in ways that made the work of the conference more complex. News of large offshore oil discoveries on the Brazilian continental shelf, for example, heightened the dispute between coastal and noncoastal states over rights to shelf resources. Similarly, news of moves by the United States to incorporate the Mariana Islands into the American commonwealth raised likely points of dispute within the conference over both colonialism and the regime of islands. The increasing intensity of offshore oil exploration and exploitation, especially by developed states concerned with security of supply, raised potentially serious delimitation disputes between Norway and the Soviet Union, Britain and France, and Britain and Ireland (over the islet of Rockall), and brought the simmering Greek-Turkish dispute in the Aegean close to boiling yet again.

All things considered, however, the intersession period passed remarkably smoothly. When the delegates reassembled in Geneva in March 1975, they were able to pick up the threads from Caracas in largely the same pattern as they had left them. The climate and state of the negotiations had neither improved nor deteriorated notably during the intervening seven months, and the delegates were able to proceed without hindrance to fulfill the promise they had made at Caracas to get down to negotiations.

THE GENEVA SESSION

Like the Caracas session, the Geneva session did not evolve in the way most observers had expected. The goal of the session was the same as that for Caracas, namely to negotiate agreed draft articles covering the mandates of the three committees. These negotiated draft articles would form a single negotiating text to replace the numerous alternative texts still remaining from the Caracas session. This would constitute the necessary breakthrough from a situation of diverse national and group positions on key issues to a situation where the most crucial conflicts had been largely resolved. At Geneva the Caracas promise to move straight into serious negotiations was kept, and from the beginning the focus of activity moved away from the main committees and into a host of working groups, negotiating groups, consultative groups, and other informal structures. Very few formal proposals were submitted to the conference, and very little time was spent making statements of position for the record. Despite this promising start, however, it became clear by the middle of the session that little progress toward agreement was being made anywhere, and consequently that negotiated draft articles were no longer a feasible objective for the session.

As a result, the conference was faced with two closely related problems: first, it had to reexamine its methods of work, and second, it had to find some way of putting an acceptable face on what threatened to be a session that produced nothing except further clarification of positions. The problem of methods was very serious, for none of the conference's chosen approaches to negotiation had succeeded in breaking states away from their declared positions. Regional, political, and issue-specific negotiating groups were tending to harden conflicts by providing positional bastions difficult to surrender for those within and unacceptable to those without. Setting up informal consultative groups on specific issues, as was tried in C.II at Geneva, did not overcome the unwillingness of these groups to be the first to abandon their positions. The so-called wise men approach, represented most prominently by the Evensen group, was not proving acceptable to the many states not included in its membership, and expansion of the membership to overcome this resulted in further slowing down of the group's already laborious drafting process. In addition, many participants expressed reservations about the drafts produced by the group as compromise texts. For all these reasons the conference interest groups failed to fulfill the crystallizing role that had been their objective. Attempts in the three committees to work as an informal group of the whole simply bogged down in the logistical difficulties of negotiating within a large and varied group. None of these approaches had succeeded in breaking through the barrier of multiple positions contained in the alternative texts.

The problem of giving the session a good appearance was equally urgent, because if it was seen to have stalled in a deadlock at Geneva, then the momentum of the international negotiations might be irretrievably broken. States would be tempted to take unilateral or regional action on the law of the sea, and much that was only possible through global agreement, and that most states wanted and had worked for, would be lost.

A solution to both these problems was found in an idea arising from the methods of work used in the informal working group of C.I and in other informal bodies, especially the Evensen group. The practice in these groups was for the chairman to prepare informal single negotiating texts (texts without alternative drafts for each article), using his own initiative to formulate compromise texts for points where the discussion in the group revealed differences of opinion. The group would then discuss the chairman's text, and he would prepare revisions if dissatisfaction still existed. Ideally this process would continue until a single draft text had been agreed upon by all the participants. The advantage of the procedure was that it gave the initiative in formulating compromises to the chairman, and thereby enabled the negotiations to overcome the unwillingness of delegates to abandon their own positions. Its disadvantage was its slowness, for even six revisions of an Evensen group draft on fisheries within the economic zone had not, after an extended period of time, resulted in a text fully accepted by the members of the group.

At the end of its fifth week, the conference decided to adopt this method as a means of providing itself with the single negotiating texts it needed very badly, but which its committees were unable to produce by negotiating. It commissioned the chairmen of the three committees to prepare informal single negotiating texts covering the entire mandates of their respective committees. This move was unpopular in many quarters, among other reasons because of the personalities of the three chairmen, who had not been appointed with this type of responsibility in mind. Its acceptance by consensus can be seen largely as an act of desperation reflecting the conference's inability to make progress by any other route. At the time this decision was taken, the plan was for the chairmen to present their texts before the end of the session, thereby enabling negotiation based on them to begin. But within a week it was clear to all present that the texts would be extremely controversial when they appeared and that a massive adverse reaction to them would shatter the facade of accomplishment that the session was trying to present. Consequently, the texts were not released until the last minutes of the session, giving nobody a chance to voice reaction to them in the spotlight of conference publicity. A side effect of this delay was to stifle informal negotiations for the last three weeks of the session. There seemed little point in continuing to negotiate under the shadow of the impending texts, and so most energies were turned toward catching the ear of the chairmen themselves, one of whom discreetly left for home a day or two before his text was made public.

The single negotiating texts, it was hoped, would provide the necessary methodological breakthrough for the conference by giving the delegates a coherent set of draft articles covering all issues and reflecting possible compromises for all key areas of disagreement. The texts gave the superficial impression of fulfilling the goal of the three committees to produce agreed draft articles covering their mandates. In fact, this was very far from being the case. Agreed draft articles would have constituted a single negotiated text, and as such would have meant that the conference had broken the back of its negotiating problems and was within sight of the end of its work. The texts actually produced did not commit anyone to anything, and were only products of negotiating in the most limited and preliminary sense. At best, they started the conference off on a cycle of negotiation and revision, which if the experience of the much smaller Evensen group was anything to go by, would take a long time before arriving at a single negotiated text.

The texts succeeded in giving the Geneva session an appearance of accomplishment, but they left the negotiations in much the same vulnerable condition as they had been at the end of the Caracas session. The breakthrough at Geneva was procedural rather than substantive, and this time the negotiations were to be suspended for a whole year, during which states would be free to abandon what little agreement did exist within the conference.

The closed, negotiating nature of the Geneva session, as opposed to the open, declaratory nature of the Caracas session, meant that the main weight of evidence at Geneva on developments in seabed politics came from informal documents emanating from various working and negotiating groups. The approach used in this chapter will therefore be to relate the contents of these documents to the trends already established at Caracas and in the SBC, and also to relate to these trends the single negotiating texts produced by the three committee chairmen.

Continental Shelf Questions

At Geneva, the only activity concerning continental shelf limits took place in the Evensen group. A small ad hoc group of margin states, mostly developed, prepared seven draft articles containing the 200-mile-plus-margin formula along with provisions for delineating precisely the outer edge of the margin according to geological and distance criteria. These articles were incorporated almost verbatim into the fourth revision of the Evensen negotiating text on the continental shelf, and were intended by the margin states to put an end to fears that an imprecise statement like "the edge of the margin" might leave open possibilities for unexpected claims to the international seabed area. These steps toward a precise definition for the margin were seen by some of the leading margin states as an important and necessary development. Despite this, the single negotiating text for C.II (WP.8/Part II) adopted only the 200-mile-plus-margin formula and contained no provisions for delineating the outer edge of the margin.

On the question of rights, there were a number of developments at Geneva, mostly in the context of drafts on the economic zone. The idea of sharing revenue with the seabed authority received attention from two quarters. On the one hand, the landlocked and geographically disadvantaged states moved away from their proposal that revenue from the whole shelf should be shared. In their group draft on the economic zone (May 2 revision) they dropped references to this that had appeared in earlier drafts and in proposals made at Caracas. Since this idea had no significant support outside the landlocked and geographically disadvantaged group, it could probably be considered a dead letter by the end of the Geneva session. On the other hand, the key compromise of sharing with the authority from the margin beyond 200 miles developed into a much more likely option than it had been before. The Evensen group included articles covering the idea in its continental shelf text, and these articles appeared almost verbatim in WP.8/Part II. In addition Canada, the United States, and Britain all accepted the principle of sharing beyond 200 miles, and indicated their willingness to begin negotiations in search of a formula. In the course of discussions on this, a number of difficulties of operationalization became clear, including the danger of stifling development on the outer margin by setting the sharing level too high, and the question of whether poor developing countries like Bangladesh should be required to share from their outer margins.

On the question of rights for landlocked and geographically disadvantaged states to share in shelf resources, the situation remained much as it had been at Caracas. Only the economic zone draft of the landlocked and geographically disadvantaged group contained articles covering this, and they expanded on earlier proposals by outlining joint ventures and other aspects of putting such sharing into practice. The landlocked and geographically disadvantaged states showed no signs of abandoning this position at Geneva despite the complete absence of support from any other quarter. Reflecting majority opinion, WP.8/Part II contained no reference to sharing of shelf resources with noncoastal states. To all appearances, the landlocked and geographically disadvantaged states had lost this battle at the 1958 Law of the Sea Conference, when they did not contest the Continental Shelf Convention.

The balance of opinion on sharing fisheries resources in the economic zone with noncoastal states also remained much the same at Geneva as it had been at Caracas, and continued to lack generally accepted, uniformly defined, provisions for implementation. The sixth revision of the Evensen group draft on the economic zone made provision for such sharing, as did the landlocked and geographically disadvantaged group draft, and also a draft by the Group of 77 (May 1) which, like the Evensen draft, did not represent a complete consensus of the group. The Group of 77 draft was evidence of an intensive effort by the Group to find an accommodation with the noncoastal states, and consequently was more favorable to those states than the Evensen text. This reflected the large number of noncoastal states within the Group of 77 and the relative underrepresentation of such states in the Evensen group. In the event, WP.8/Part II enshrined this key compromise in terms somewhere between the Evensen and Group of 77 drafts.

An interesting addition to the debate on rights within the limits of national jurisdiction came from Ecuador, a leading territorialist state, in the form of a proposal on a 200-mile territorial sea submitted to the plenary meeting of C.II.[2] The proposal spelled out the territorialist alternative to the economic zone, and was notable for the absence of any reference to resource sharing rights for noncoastal states. In effect it defined the extreme opposite view to that of the landlocked and geographically disadvantaged group. Although it was not a proposal of the territorialist group, it did indicate the extent of that group's support. Among those speaking favorably in response to it were several nonmembers of the group including Guinea-Bissau, the Malagasy Republic, North Korea, China, and Albania. The territorialist view did not find favor in WP.8/Part II, which contained articles outlining a 12-mile territorial sea and a 200-mile economic zone.

On the continental shelf questions, then, the Geneva session strengthened the trend toward compromise among the margin, flat 200-mile, and noncoastal positions that was already becoming evident at Caracas. The two key compromises of sharing with the authority from the margin beyond 200 miles, and sharing with noncoastal states from fisheries resources in the economic zone, both gained ground and achieved the stability of incorporation into WP.8/Part II. Thus, despite the continued campaign of the landlocked and geographically disadvantaged states on the question of sharing shelf resources, the Geneva session moved a long way toward closing the gaps on continental shelf issues between different schools and groups on both questions of limits and questions of rights. As at Caracas, however, the whole economic zone package of which the continental shelf was a part continued to rest on very unstable ground. Little progress toward agreement on straits, scientific research, or pollution was made, and the compromises suggested in the single unified texts of the committee chairmen seemed unlikely to paper over successfully the deep rifts on these issues. The outlook on fisheries within the zone was somewhat improved by the Evensen group's work, most of which was incorporated into WP.8/Part II, but even this was still a long way from conference-wide agreement.

International Seabed Regime and Machinery Questions

In C.I, the focus of activity moved straight into the working group on Article 9 and the conditions of exploitation that had been set up at the end of the Caracas session. The working group concentrated its attention on the conditions of exploitation, taking as the basis of its work the four papers on that subject submitted at Caracas plus a fifth paper submitted by the Soviet Union near the beginning of the Geneva session.[3]

The Soviet paper was, with a few exceptions, generally in line with those of the other developed states in that its primary emphasis was on the rights and duties of operators, with stress laid on states as the entities with which the authority would enter into contracts. Three aspects of the paper stood out as distinctive. First, the authority would be allowed to reserve a set percentage of the seabed areas open to exploitation to exploit "with its own means." This would, in effect, divide the seabed into two separate regime areas, one being exploited by states in conformity with the authority's rules and contracts, and the other being exploited by the authority under whatever system it wished, or was able, to implement. Second, the Soviet paper displayed strong concern that seabed operations not interfere with other uses of the seas. This reflected the strategic concern of the Soviet Union that nothing happen to jeopardize its geographically precarious access to the high seas. Third, the paper contained much more generous provisions for the suspension of operator's rights than any of the Western papers, and indeed on this aspect tended to be more in line with the Group of 77's approach than with that of the other developed states, although that tendency was substantially moderated by Soviet agreement with the Western states on the need for dispute settlement procedures. The Soviet attitude on this appears to have been shaped by a lack of enthusiasm for any near future Soviet venture into nodule mining, and also by an understandable lack of concern for the capitalist sensitivities of Western mining companies.

With the Soviet paper completing the set, the working group continued its negotiations on the conditions of exploitation. For the first few weeks, most delegates felt that the discussions were constructive, and there was considerable optimism about the committee's work. The idea of joint ventures enjoyed increasing popularity as a means of overcoming the "who shall exploit?" dilemma, with both the United States and Britain warming to it. The chairman of the group, C. W. Pinto, submitted an informational paper on the subject that was successful in encouraging further discussions on joint venture alternatives.[4] A measure of the progress being made in the working group was that the chairman felt able to submit, during the fourth week of the session, a single negotiating text (CP/cab.12) on the conditions of exploitation that used the joint venture theme in an attempt to provide a unified basis for discussion.

From the midpoint of the session onward, however, the work of C.I suffered a number of setbacks. The extremely able chairman of the working group succeeded in making CP/cab.12 the center of the negotiations and managed to gather considerable support for many of the compromises suggested in his single text. But his text contained a provision, similar to that in the Soviet paper, that the international seabed area should be divided into two separate regimes, one exploited by state contractors and the other by the authority using its own means. The arrangement was given a twist suggested by the United States, that applicants for contracts would have to submit two proposed mine sites, one of which they would receive rights to and the other of which would revert to the authority to be exploited by it in any manner it might determine within the limits of the convention. The idea of somehow dividing the seabed into state and authority areas, however, proved as unacceptable to the Group of 77 in this guise as it had in earlier mixed regime formulations. Consequently, after a lengthy and heated internal debate on cab.12, the Group decided to reject the document, and particularly the divided area provisions, as an unsuccessful attempt to patch together conflicting proposals.[5]

Since the United States in particular had pinned its hopes to cab.12 as an acceptable means of pursuing negotiations in C.I, rejection of the document by the Group of 77 precipitated a crisis in the committee's work. Bitterness prevailed in many quarters over the collapse of cab.12, and the chairman of the working group came under criticism from the more radical members of the Group of 77 for having conceded too much to the advanced technology states in the first place. For a few days, the negotiations in C.I appeared to be on the verge of a complete breakdown.

Despite this situation, the chairman began consultation with a view to producing a revised version of cab.12. At the same time, he circulated a draft text on the structures, functions, and powers of the international machinery, and began consultations aimed at producing a draft on this for submission to the chairman of C.I in his capacity as official drafter of the single negotiating text for the committee.[6] During these consultations, controversial aspects of the structures, functions, and powers of the authority became an increasingly prominent feature of the negotiations, largely as a result of U.S. concern over the direction in which the discussion of conditions of exploitation was going.

The resurfacing of structures and powers questions, which had been dormant since 1972, marked the failure of the attempt to tackle the C.I negotiating problem on the basis of the key issues of who exploits and the conditions of exploitation. It meant that the central point of dispute had been chased yet again into another form, in that detailed discussions of the conditions of exploitation simply raised more and more questions about the nature of the authority that would be implementing those conditions in practice. Wherever powers of decision were vested in the authority, the structure of the authority, particularly the nature of its decision-making and its internal distribution of power, became matters of great relevance. This tendency was amplified by the deep distrust between the two sides in the negotiations, which created a condition in which everything had to be explicit before it could be accepted.

In the event, C.I staged a quite useless plenary debate on the structure, functions, and powers of the machinery. This accomplished nothing except the reiteration of well-known and mostly sketchy statements of position. This debate gave notice that the other aspects of C.I's mandate were not going to fall smoothly into place even if some preliminary agreement on conditions of exploitation could be hammered out. Consequently C.I no longer faced the possibility of a relatively rapid conclusion to its work, but instead confronted the necessity to treat each aspect of its mandate as a major item probably requiring not only negotiation in its own right but also successive renegotiation in the light of what was achieved on related aspects. In other words, there was no key to the C.I negotiations, but only a succession of locks each having to be picked in turn.

During the last weeks of the session, C.I did not succumb to the same degree as the other two committees to the intense lobbying of the committee chairman on the single negotiating texts. This was because consultations on the working group chairman's two texts were still in progress, and delegates assumed that revised versions of these two texts would constitute the single negotiating text for C.I.[7] On the basis of these consultations, the working group chairman prepared revisions of both his texts and, during the last week of the session, submitted them to the committee chairman. Although there was insufficient time for a full reaction to the two texts to develop, such evidence as became available indicated that neither side was very happy with them, and that the level of commitment behind them was consequently rather low.[8] On the positive side, however, the contents of the texts revealed substantial movement by both sides on a number of important issues, and this made the two texts a significant advance in the committee's work even if they were only acceptable as a basis for further negotiation.

One last crisis broke upon the committee in the final week of the session when it became clear that the committee chairman, P. B. Engo, was not going to accept the working group chairman's texts without imposing some alterations of his own. Although the working group chairman's revised text on conditions of exploitation (CP/cab.12/Rev.1) was placed verbatim in the C.I single negotiating text (WP.8/Part I), his text on the machinery, which had received much less public circulation than cab.12/Rev.1, was altered in several places. These alterations worked for and against both sides, although in total they favored the developing countries.[9] What was important was the possibility that they would endanger the already fragile balance of concessions reached in the working group. Since the committee chairman had the last word in the matter, his revisions appeared in WP.8/Part I.

The importance of WP.8/Part I as a whole was not that it represented an accepted compromise that would merely require polishing at the next session, but that it provided a common basis for discussion around which the negotiations at the next session could be focused. The experience of the Evensen group, however, indicated that several major rounds of negotiation and revision would be necessary before anything like an accepted single text could emerge. For this reason, a detailed summary of the text's provisions is not necessary in this book. What is required is an analysis of how the text related to the existing alignments in C.I, since it is this aspect that will determine the nature of the negotiations at the next session.

Somewhat surprisingly, WP.8/Part I turned out to be more favorable to the interests of the advanced technology states than might have been expected. To some extent, the whole text can usefully be viewed as an exercise in weaving U.S. demands into the verbal framework set by the Group of 77, and this outcome illustrated the increasing influence that the passage of time gave the United States on this issue because of its strengthening unilateral option. On many, though by no means all, of the issues under dispute in C.I, the compromises contained in WP.8/Part I reflected demands made by the United States at both Caracas and Geneva.

On the issue of minimizing adverse economic effects on developing country metal exporters, the text included the principle of minimization or avoidance, but the relevant powers were vested in the council. Thus, while the form of the Group of 77's demands on this issue was preserved in an economic planning commission, the commission itself had only recommendatory power to the council and was therefore beyond the influence of the developing country majority.

On the issue of transfer of technology, the authority was given only rather general obligations that would not threaten the proprietary information of private companies. This approach, however, did not accord with that taken in the text for C.III, which imposed obligations on the authority more in line with those envisaged by the Group of 77. This was one of a number of contradictions among the three parts of WP.8.

On the high seas superjacent to the international seabed area, the text made clear the legal separateness of the two areas, and thereby accommodated the demands of the maritime powers.

On the structure of the authority, the text accommodated almost every important point demanded by the U.S. delegate in a speech to C.I on April 28. Voting in the assembly was to follow the Law of the Sea Conference format in that a two-thirds majority must include a simple majority of the members, and also in that powers to delay voting were given to groups mustering support from at least one-third of the members. In return for this, the United States conceded some policy-making powers to the assembly. The structure of the council gave weight to the interests of states involved in nodule mining, and voting was arranged on a two-thirds-plus-one majority basis, which meant that 12 states could form a blocking group. Since the proposed balance of membership would ensure the presence of more than 12 developed states, the council in this form would give protection to the developed state minority. (A ratio of membership even more favorable to developed states appeared in the working group text, but this was altered by the committee chairman, making this a likely point for renegotiation at the next session.[10]) In terms of powers, WP.8/Part I preserved the rhetoric of the assembly as the supreme organ of the authority, but vested most of the important powers and functions in the council. The chairman of C.I also added a key interpretation to the text by saying that in the case of conflict between the two organs the assembly could not overrule the council on a matter within the council's jurisdiction.[11]

The text provided for strong machinery for the compulsory settlement of disputes, including a powerful tribunal, and although it also provided for an enterprise, this was to be largely subordinate to the council. The operations commission, a body favored by the United States, was not included as a result of the chairman's cuts in the working group chairman's text. One of the few areas of contention decided in favor of the Group of 77 was on the issue of state contributions toward the running costs of the authority, where powers were given to the assembly. The demands of the United States and others that the convention should become active provisionally, in order to prevent delay in implementation of a nodule mining regime, were also provided for.

On the conditions of exploitation, the nine disagreements that arose at Caracas were handled in WP.8/Part I (cab.12/Rev.1) as follows.

(1) A compromise was devised in which both the area and its resources were declared the common heritage of mankind, but in which only the rights in the resources were vested in the authority. This again illustrated the marriage of Group of 77 terminology with advanced technology state demands on substance.

(2) The question of the functions of the authority was basically decided in favor of the Group of 77's position, in that scientific research as well as processing and marketing functions were included. This was, however, moderated by the provision that the authority did not have exclusive right to conduct scientific research.

(3) The question of discretionary powers for the authority was not settled clearly, but the thrust of the compromise favored the advanced technology states and appeared adequate to compensate them for the rejection of their dual area schemes. Operators were to choose mine sites, but only within areas opened by the authority. Contracts for activities appeared to be nearly automatic if the applicant was properly qualified, with only a very few minor areas of discretion left to the authority as grounds for refusal. Furthermore, operators were given priority in award of subsequent stages of operation, and the authority was required to give priority to integrated contracts for the first contracts let up to a fixed (but not specified) number. The nature of the council improved the position of the advanced technology states in this regard, and as an additional bonus to them the council was required to open up ten mine sites to exploitation by joint venture with states or their nationals "as early as practicable after [the] coming into force" of the convention.

(4) The transfer of rights from one operator to another was nominally under the control of the authority, but the authority could not withhold consent if reasonable, specified conditions were fulfilled.

(5) Financial arrangements were not completely worked out in WP.8/Part I, but inasmuch as they were, the text appeared to offer a fairly balanced compromise. Although the authority would retain "direct and effective fiscal and administrative control," the operator was guaranteed return of his costs out of the proceeds, and also a share of the profits to be specified in the contract.

(6) The right of the authority to revise contracts was not developed in detail in the text, even though a heading providing for such powers was included. In another section, however, operators were guaranteed security of tenure and freedom from alteration or suspension of the contract except for "gross and persistent violations" of rules.

(7) to (9) Production controls, settlement of disputes, and transfer of technology were all dealt with outside the conditions of exploitation, and have been discussed above.

Given the nature of these compromises in relation to the positions taken by the Group of 77 and the advanced technology states in the SBC and at Caracas, it is possible to offer some speculations about the nature of the next session in C.I.

For the advanced technology states, the Geneva session produced mixed results and mixed feelings. While these states did quite well at the beginning of the session in cab.12, they suffered setbacks in cab.12/Rev.1, and also lost some of the gains made in the working group text on the international machinery. In terms of the positions prior to the Geneva session, however, the advanced technology states appear to have made considerable gains. The single negotiating text for C.I represented a substantial movement away from the rigid confrontation lines in the SBC and at Caracas, and much of this movement was in directions favorable to these states. On this basis it seems not unlikely that the advanced technology states would accept WP.8/Part I as a starting point for further negotiations and would concentrate on revising the parts that remain vague, incomplete, or unacceptable.

The major qualification to this scenario is the assumption that the United States will retain its commitment to the negotiations. Some sections of the U.S. delegation were reportedly very unhappy about the direction of the negotiations in C.I,[12] and U.S. commitment seems likely to depend on the outcome of another round of bureaucratic politics in Washington. This outcome is made uncertain by the weakening effect that the slow pace of the negotiations had on the pro-conference forces in the United States. Should the U.S. position in C.I harden for domestic reasons, then the movement toward compromise at Geneva might easily be rejected as still tending toward a regime unacceptable to the United States. Perhaps the key factor determining the U.S. position on the C.I negotiations will be the hope of obtaining a package deal covering the sensitive question of passage through straits.

The Group of 77, for its part, seems much more likely than the advanced technology states to face a major crisis over the single negotiating text. The compromises made in WP.8/Part I were incompatible with many of the positions taken by the Group in the SBC and at Caracas, and unless it can unite in condemnation of the text, it seems likely to fragment in the process of producing a response. The Group's record on Article 9 and the conditions of exploitation demonstrates the difficulty it has had in reaching an internal consensus and in negotiating on that consensus with external states. Its response to cab.12 illustrates the even greater difficulty it has had in coming to terms with a document produced outside its ranks, and offers no grounds for optimism about its ability to retain its unity in dealing with WP.8/Part I. The Group's debate on cab.12 revealed severe internal divisions that are almost certain to become prominent if WP.8/Part I is made the basis for negotiation at the next session. The individual members of the Group are thus likely to be faced with a choice between accepting the single text as a basis for negotiation, but at a cost of abandoning Group of 77 unity, or rejecting the text and maintaining Group unity, but at a cost of jeopardizing the negotiations. Although many of the more moderate members could probably accept the text as a basis for negotiation, the laboriously achieved and symbolically important unity of the Group of 77 could become an impediment to the process of compromise in C.I.

Another possibility (one stressed by Edward Miles[13]) is that the commitment of both sides to the creation of an international seabed regime and machinery will decline sharply because neither will accept the range of solutions possible in any agreement between them. In this sense, the movement at Geneva represented by WP.8/Part I would merely have served to illustrate to both sides that accommodation was not possible. The implication of this scenario is that the negotiations in C.I would collapse, possibly taking with them the rest of the conference. Should C.I appear to be jeopardizing or delaying agreement on the economic zone it seems likely that pressure would rise to separate it from the rest of the agenda. Such an action would have complex ramifications for the package deal approach favored by the United States.

Assuming that the negotiations were not on the verge of breakdown, the crucial question about the single negotiating text was whether or not it constituted the breakthrough necessary to extract C.I from the morass of factors tending to stretch its work out interminably. At the end of the Geneva session, the answer to that question appeared to be no. Although the text did solve a tricky procedural problem, the path to progress it opened up was a long one, and the mere presentation of the text did nothing to hide the severe differences on almost every matter of substance before the committee. The glacial pace of the conference's progress toward agreement at Geneva effectively killed the whole idea of a breakthrough, and in this sense WP.8/Part I simply confirmed the realization that the conference was going to take a long time to finish its work.

As a procedural device, the single negotiating text had the potential to circumvent some of the slowing effects of factors like the large number of delegates and issues, the consensus procedure, and the polarized alignment pattern within the committee. Unfortunately, these gains were won at the cost of the slow method of work represented by the text itself, leaving the balance only slightly, if at all, improved from what it had been before. There were some changes in the other factors slowing down the committee's work, but on balance these also continued to form an insurmountable barrier to rapid progress.

The developed/developing polarization remained the order of the day at Geneva, but there were a number of signs that these hard lines were beginning to crumble. Although the Western advanced technology states were more united than at Caracas, a number of developing countries had hopes that the Soviet Union could be detached from the developed states on the basis of its different concerns (the Soviet Union was an exporter of nodule metals rather than an importer) and its dislike for Western private companies. The Group of 77 also appeared less united than at Caracas, with the landlocked and geographically disadvantaged group fielding separate proposals, and with obvious rifts of opinion on subjects like joint ventures. The Group's need to deal with the single text seemed likely to exacerbate this disunity at the next session. This process of loosening alignments might facilitate the negotiations by strengthening the center ground so long occupied only by Canada and one or two other uncommitted states.

Initiative-taking by states remained static up to the Geneva session, with the Metcalf bill still present only in the background of U.S. ocean politics. The failure of the conference to complete its work by the 1975 deadline seemed likely to reopen this issue in the United States, because the executive had based its case for internal restraint very heavily on the conference having an acceptable deadline for completion. Should a Hard Minerals Bill become active in the Congress, tensions in C.I would be raised enormously.

Developments in the nodule mining industry continued to evolve in a rather quiet way, with steady progress toward an operational technology and further formation of consortiums, some without an American company. The revelation that Howard Hughes' Summa Corporation had been using nodule mining as a front for government intelligence work came right in the middle of the Geneva session, but caused surprisingly little stir. The company insisted that it was still in the nodule business, and the main impact of the revelations fell on the debate over scientific research, where developed states took a verbal beating on the issue of freedom for such activities. Rather curiously, the developing countries made little comment on the nodule industry and its works, apparently feeling that it was beneath their dignity to comment on entities below the level of governments. Their restraint or indifference on this subject minimized a potentially severe source of friction in the debates.

Several problems relating to genuine complexities in the issues before C.I also continued to be impediments at Geneva.

The influence of the global commodity debate on the work of C.I became more explicit, and was much more openly recognized and discussed by delegates. This tended to increase the symbolic value of the international seabed regime and machinery by casting it in the role of the thin edge of the wedge of the new economic order. This symbolic aspect increased the already formidable difficulty of negotiating compromises in C.I, but helped to restore some of the priority that C.I had lost to C.II. Moves toward agreement on the shelf and fisheries in C.II also helped C.I to resume a more prominent role in the conference.

The politicization of information continued in much the same vein at Geneva as it had done at Caracas, although the move away from discussion of the conditions of exploitation tended to reduce its importance. A report of the Secretary-General on the economic implications of seabed mining in the international area tended to favor the view of the advanced technology countries, while the view of the Group of 77 was reinforced with further UNCTAD reports.[14]

The possibility of a major dispute over high seas fishing never materialized at Geneva, and this problem appears to have died as a threat to C.I because the developing countries realize that they stand to lose more than they would gain if they pushed it. The end of this as a problem also marks a setback for the idealist critics of the conference, including Arvid Pardo, who wanted to expand the common heritage concept to include ocean as well as seabed resources.

The problem of the regime of islands was not much discussed at Geneva, and so remained as a difficulty confronting future sessions. WP.8/Part II made relatively generous provision for archipelagic states, but denied an economic zone and continental shelf to islands incapable of sustaining human habitation or an economic life of their own. A regime along these lines would definitely have serious implications for the resource monopoly of the seabed authority.

The problem of transfer of technology became more serious in C.III with the adoption by the whole Group of 77 of the text presented by 18 developing countries at Caracas. A number of developing countries chided the developed states for not making any proposals on the transfer of technology, and both the United States and the Soviet Union expressed disapproval of the Group of 77's text.[15] Furthermore, the single negotiating text for C.III included some of the more controversial demands from the Group of 77's proposal relating to matters of proprietary information. Balancing this was the much milder line taken in WP.8/Part I, and the general drift in C.I discussions to leave the problem to be resolved within the contracts to be negotiated by the authority. The increasing disparity between the work of the two committees on this issue remained to be resolved by subsequent sessions.

One additional problem that arose to disrupt C.I's work at Geneva was the personality clash between the committee chairman, P. B. Engo, and the chairman of the working group and the informal sessions, C. W. Pinto. This conflict had been visible at Caracas, but had not become prominent because of the concentration of the committee's work in the informal sessions chaired by Pinto. The same would have been true at Geneva except for the thrusting of responsibility onto the committee chairman by the decision to produce single negotiating texts. This highlighted the different working methods and personalities of the two men, and made public the competition between them for leadership of the committee's work. Continuation of the dispute would divide loyalties in the committee and add an unnecessary burden to its already heavy load.

The Groups

Because of the informal nature of the negotiations at Geneva, the activities of negotiating groups were much harder to document than at Caracas. Only a few formal proposals indicating cosponsoring were made, the bulk of the documentation consisting of informal proposals. These informal documents frequently indicated a much weaker level of support from the states cosponsoring them, or involved in their drafting, than would have been the case for formally submitted documents. The session nonetheless produced sufficient evidence for an assessment of group activity.

The Group of 77 turned in a mixed performance at Geneva, with some gains and some losses in strength. As previously, it maintained its very powerful symbolic position at the conference for the developing countries, in spite of the many disappointments with its functioning in the negotiations. In C.I, the Group appeared weaker than at Caracas for the reasons described above.

In C.II, the Group of 77 made a substantial advance over its limited role at Caracas by beginning work on a position paper on the economic zone. During the session this work reached a stage where a preliminary draft (not, however, representing a consensus of the Group) could be made public. The emphasis of this draft revealed a major effort within the Group to come to terms with the deep split between coastal and noncoastal states on economic zone issues. Although this development marked a great improvement for the Group in C.II, it did not in any sense indicate agreement within the Group on the economic zone. Both the landlocked and geographically disadvantaged, and the territorialist alternatives, continued to be represented in separate documents.

In C.III, the Group of 77's role stayed on much the same level it had reached at Caracas. The Group made some improvements on its scientific research proposals (C.3/L.13 Revs.1 and 2) but continued to be plagued with dissidents among the landlocked and geographically disadvantaged states, as at Caracas (C.3/L.28, containing amendments to C.3/L.19), and also among more moderate coastal state members, as exemplified by the compromise draft on scientific research submitted by Colombia, Mexico, El Salvador, and Nigeria (C.3/L.29). It made a notable advance with the Group-wide consensus on the transfer of technology (C.3/L.12 Rev.1), and did some preliminary work on the problem of marine pollution (Working Group/Paper No.4/Rev.2).

In general, the Group of 77 became a more important element of the Conference at Geneva than it had been before. This was partly due to the Group's better performance in consensus-building in C.II and C.III, and partly due to the decline, inactivity, impermanence, or weakness of the other main negotiating groups representing particular as opposed to conference interests. The Group's vital role in the new economic order debate outside the Law of the Sea Conference appeared to account for much of its increased prestige and activity at Geneva, and it was subject to the strength of the optimistic "time is on our side" mentality among many of its member delegations.

A logical result of the Group of 77's greater prominence was a lesser role for its constituent regional groups. The African, Latin American, and Arab groups all held regular meetings at Geneva, but most of these appeared to relate to the Group of 77's work, and no proposals were forthcoming from any of them. A group of Central American states also held a number of meetings, but likewise made no proposals. There was some evidence that the formidable unity of the African group based on its 1973 and 1974 declarations was beginning to dissolve due to the difficulty of operationalizing the compromise between coastal and noncoastal states on sharing living resources in the economic zone. The relatively unified and active performance of the landlocked and geographically disadvantaged state group at Geneva underlined this decline.

Among the developed states, only the Soviet group remained active as a regional group, and even it held few meetings and produced only one proposal (C.3/L.26) on scientific research. The Soviet Union sponsored two of its major proposals alone (C.1/L.12, and C.3/L.25). The EEC group held some meetings but was very fragmented and produced no proposals.

Much of the explanation for the relative quiescence of the regional groups at Geneva lies in the transfer of activity to other types of groups. The Group of 77 as a super-regional political group was one of these types; the second was ad hoc (and also, though to a much lesser extent, more permanent) issue-specific groups; and the third was conference interest groups, in particular the Evensen group.

Among the more permanent issue-specific groups, the three singled out in this study as of particular relevance to seabed politics followed much the same pattern of activity as at Caracas. The coastal state group remained a forum for contact, but met very infrequently and played no part in the negotiations, most of its members preferring to concentrate their activity in the Evensen group.

Predictably, the territorialist group became much more visible at Geneva. It held frequent meetings and adopted the strategy of making its presence felt by raising the level of visibility of its activity, for example by announcing its meetings in public. The group did not demonstrate sufficient unity to sponsor any documents, but a number of its Latin American core members could be seen to work together in the C.II negotiations, and something of the group's full support could be gathered from the more than one dozen states that responded favorably to Ecuador's formal proposal on the territorial sea (C.2/L.88).[16] The full emergence of this group at Geneva was an ominous indication of the declining momentum of the international negotiations. Since the territorialist position represented the most extreme coastal state view, it contained almost no room for compromise with maritime and noncoastal interests, and was therefore fundamentally opposed to the basic compromising purpose of the conference.

As at Caracas, the landlocked and geographically disadvantaged group was by far the largest and most active of the issue-specific groups relating to seabed politics. Its membership was up to nearly 50 states, it held frequent meetings, and it produced four documents. In C.I and C.III, which covered areas peripheral to the group's central interests, it either maintained or strengthened its position. In C.I, the group as a whole submitted two formal proposals. One, relating to representation in the organs of the seabed authority, demanded two-fifths representation (this was proportional to their numbers in the UN) for landlocked and geographically disadvantaged states (L.14). The other outlined a system of revenue distribution favorable to landlocked and geographically disadvantaged states (L.13). These documents were the first to breach seriously the developed/developing alignment in C.I. In C.III, the group maintained its own distinctive position on scientific research as described above. In C.II, the group took the same course as the Group of 77 and the Evensen group, and produced its own draft proposal on the economic zone. As discussed above, this document contained draft articles outlining a sharing system for noncoastal states in both the living and nonliving resources of the zone, and was based on the idea that the zone belonged to the continental mass and not just to the coastal state.

Despite all this, the group's position after Geneva was weaker than it had been after Caracas. Five factors additional to the group's underlying weakness (no unilateral option for its members) accounted for this. First, the Group of 77 made a number of moves toward reabsorbing its landlocked and geographically disadvantaged dissidents, the most obvious of which was its draft on the economic zone, which was in large part concerned with the problems of noncoastal states. Their effect was to open a gap between developed and developing members of the group. At Geneva this gap was marked by the frequent meetings of a group of landlocked and geographically disadvantaged states of the Group of 77—in effect a subgroup of the 77—the existence of which indicated a trend toward dissolution of the group as an independent entity crossing regional/political lines.

Second, the group was divided along landlocked/other lines by the issue of right of transit, which was a major preoccupation for the landlocked but of little interest to the nonlandlocked. Third, there was considerable evidence that the leading coastal and maritime states did not consider the landlocked and geographically disadvantaged states to be an important group at the conference. The almost exclusively coastal and maritime membership of the Evensen group highlighted this attitude, and many delegates from the coastal and maritime states viewed the landlocked and geographically disadvantaged group as an obstructing nuisance or, particularly among developed country delegates, as an internal problem for the Group of 77. In this view, the demands of the landlocked and geographically disadvantaged states for equal rights to ocean resources were seen as unrealistic and overinflated, and the group was seen as potentially disruptive to the conference because of its voting power. In effect these delegates took the view that the coastal and maritime states were both the most important and the most powerful states involved in the law of the sea negotiations, and that finding agreement among them was difficult enough without the unreasonable demands of the landlocked and geographically disadvantaged states to cope with as well.

Fourth, and closely related to this political isolation, was the group's obvious lack of influence in the drafting of the single negotiating texts. None of the group's demands in relation to the seabed were included in the texts, and the major concession to the group (on sharing fishery resources in the economic zone) was a trade off largely promoted and supported by coastal states. Finally, the group's position was eroded by the poor performance of the Geneva session and the consequent weakening in the momentum of the international negotiations, on which the group was entirely dependent for its influence. As a result of these factors, the landlocked and geographically disadvantaged group, though superficially strong and active, virtually ceased to be a force in seabed politics by the end of the Geneva session. In direct terms, its only victory in relation to the seabed was the probable acceptance by the margin states of the compromise on sharing beyond 200 miles.

As at Caracas, many ad hoc issue-specific groups flourished at Geneva. At Geneva, however, they played a more prominent role because of the negotiating, rather than declaratory, nature of the session. These groups had widely varying degrees of longevity, coherence, and productivity. Some were organized around promotion of a particular interest, while others were oriented to proposing compromises among conflicting interests. In general they were better suited to the detailed level and rapid pace of the negotiations than the more permanent groups, and they partially account for the relative lack of activity among these latter groups at Geneva. Perhaps the most useful way to look at these ad hoc groups is as subgroups relating to the work of the more permanent, active groups, particularly the Group of 77 and the Evensen group. Most of them were active in C.II and C.III, and the most important one relevant to seabed politics was the group of margin states, discussed above, which was a subgroup of the Evensen group.

The Evensen group itself was unquestionably one of the key negotiating forums at Geneva, ranking with the Group of 77. While it fell a long way short of fulfilling the expectations of some observers that it would present the conference with a single negotiating text, it did play a major role as the center of activity for many important states, and was an important part of such moves toward compromises as occurred in C.II.

Nevertheless, for a number of reasons, the Evensen group was not as influential at Geneva as had been expected. Its working method proved slower in practice than had been hoped, which meant that the group's work was not sufficiently advanced at Geneva to enable it to take a conference-wide initiative in the negotiations. Partly as a result of this, there was considerable dissatisfaction among its members, with working drafts being released prematurely before full agreement on them had been reached. The group also came under criticism, particularly from developing countries, on the grounds that it represented an elitist approach to the negotiations. Efforts to meet this by expanding membership simply slowed down the group's work and increased dissatisfaction. There was no escaping the fact that the essence of the group's approach was elitist, and this created not only a destructive "ins versus outs" feeling about the group, but also pressure on its developing country members from the vigorously egalitarian ranks of the Group of 77. In addition, the group was accused of representing coastal and to a lesser extent maritime interests rather than conference interests as a whole. This was also true given the underrepresentation of landlocked and geographically disadvantaged states, and the appearance of simply a reconstituted coastal state group was enhanced by the partial shift toward coastal preoccupations of traditional maritime states like Britain and the United States.

Significantly, opinion about the Evensen group tended to divide along developed/developing lines, with developed states often stressing the importance of the group and developing states concentrating on its faults and weaknesses. The group's economic zone and continental shelf drafts were clearly very influential in the single texts, as was the method of work it used to create the texts. By the end of the Geneva session, however, it was difficult to see a clear future for the group. The single texts had preempted its primary function, and any future role it might have as a forum for compromise would depend on the way in which states reacted to them.

CONCLUSIONS

At the close of the Geneva session, it was clear that the Law of the Sea Conference was entering a new and largely unanticipated stage in its work. The delegates decided not to hold another session during 1975, thereby postponing the next session of the conference for nearly a year to March 1976. It was widely felt by delegates that at least two more working sessions would be necessary to fulfill the mandate of the conference, and given the need for a separate signing session at Caracas, this put even the most optimistic completion date some time in 1977. As a result of this decision, and also of the relatively meager results of the Geneva session, the 1975 deadline under which the conference had been operating up to the Geneva session was left in ruins.

From the perspective of the Caracas session, the collapse of the 1975 deadline should have spelled disaster for the conference. In part because of the successful production of the single texts, however, and in part because of the extraordinary and consistently underestimated resilience of the international negotiations, such a disaster, though still very possible, was by no means certain. The single texts provided a vitally needed concrete product for the

Geneva session, and on the basis of reaction, or the lack of it, to them in the weeks following the session, it appeared as if most governments were willing to support the conference for another year. Iceland succumbed to the pressure on its fisheries by declaring a 200-mile fishing zone, and Mexico claimed a 200-mile economic zone, but Canada, another likely candidate for unilateral action on fisheries, appeared to have decided against further such action.[17] The situation in the United States, where a fisheries extension bill was before Congress, appeared to have an even chance of following that of either Iceland or Canada.

One of the outcomes of the Geneva session was a need to reassess the whole problem of unilateralism in relation to the conference. In his closing address, Conference President Amerasinghe addressed two appeals to the delegates on this subject.[18] The first was from the Group of 77, expressing the Group's concern over "pronouncements made in responsible quarters" that unilateral action would be taken on exploration and exploitation of the international seabed area if the conference failed to produce a treaty. This was aimed at the United States, and the president added his own weight to the appeal that all states and their nationals should refrain from any such actions until the conference had been allowed more time. The Group of 77's action, with Amerasinghe's support, underlined the extreme sensitivity of the whole conference to the prospect of unilateral actions on deep sea nodule mining. Many developing country delegates made clear the great symbolic importance they attached to the common heritage, and few had much hesitation expressing the opinion that U.S. action on nodule mining would wreck the conference. In part such opinions were simply a way of maintaining pressure on the United States in C.I, and as such reflected the steadily strengthening American unilateral option. The threat of disruption was not an empty one, however. In the eyes of many observers and delegates it constituted the greatest danger to a comprehensive agreement on the law of the sea.

By contrast, the president did not identify himself with the second appeal, made by the landlocked and geographically disadvantaged group, that "States not . . . take any unilateral or other measures which would extend national jurisdiction beyond 12 nautical miles before the . . . Conference . . . has completed its work."[19] The source of this appeal, and the relative lack of emphasis given to it by Amerasinghe, indicated a vital shift of opinion at the conference on the subject of unilateral claims by coastal states. By the end of the Geneva session, such claims no longer appeared to pose the grave threat to the negotiations that they had during the preceding few years.

Two principal reasons lay behind the diminution of this threat. First, the abandonment of the 1975 deadline, and the unwillingness of the conference to replace it with any later date, drastically undercut the ability of the conference to act as a restraint on unilateral action. While states could reasonably be asked to exercise restraint in relation to a known length of time, they could not reasonably be asked to base their policies on a completely open-ended schedule of international negotiations. Thus, in accepting a longer view of the conference, delegates implicitly accepted a higher tolerance for unilateral action by states facing urgent situations off their coasts.

Second, the relationship between likely claims and the substance of the negotiations had also changed. The main area of concern was fisheries, particularly in the relatively small number of states (including the United States, Canada, Iceland, Norway, and Britain) that were experiencing heavy foreign fishing off their coasts. These states had all participated in the Evensen group negotiations on fisheries in the economic zone, and it seemed very likely that if they did decide to take unilateral action, they would do so in a way that would not place them outside the framework of the negotiations. The fact that substantial, if not complete, progress toward agreement on fisheries had been made at Caracas and Geneva greatly eased this problem by providing a more solid framework to which states could relate their actions. The prospect of a flood of reactive claims to 200-mile territorial seas did not seem likely given the high level of commitment to the conference method demonstrated by most developing countries, and a flood of fishery claims was made unlikely by the fact that most significant developing coastal states had already made substantial fisheries claims beyond 12 miles.

The final outcome of the conference remained, as ever, unpredictable. The chances for complete failure to agree on a full and detailed convention seemed quite equal to the chances for a limited success of some sort in the form of a partial convention covering some issues only, or a full convention of a vague and general nature.[20] The chances of full agreement on a complete and detailed law of the sea convention seemed rather small.

In any event, however, the conference seemed assured of a long future. The simple inertia of the huge investment in time and work that it represented, added to its transcendence of the 1975 deadline, seemed likely to carry it through two or three more sessions. The commitment of states to the international negotiating process has been much stronger in actions than in words, and at this late stage in the proceedings this policy seemed unlikely to be reversed.

NOTES

1. J. E. Flipse, "Deep Ocean Mining Technology and its Impact on the Law of the Sea," in Francis T. Christy et al. (eds.), Law of the Sea: Caracas and Beyond (Cambridge: Ballinger, 1975), pp. 325-32, presents an illuminating view of Deepsea Ventures' claim.

2. A/CONF.62/C.2/L.88, and A/CONF.62/C.2/SR.48.

3. A/CONF.62/C.1/L.12. This paper was added to the comparative table of the other four in C.1/CP/Working Paper no. 2/Add.1.

4. C.1/Working Paper no. 5.

5. For a more detailed discussion of the Group of 77's reactions to cab.12, see E. Miles, "An Interpretation of the Geneva Proceedings—Part I," Ocean Development and International Law Journal, forthcoming in vol. 3, no. 2 (July/August 1975).

6. Ibid.

7. Ibid.

8. Ibid.

9. For details of the alterations, see ibid.

10. Ibid.

11. P. B. Engo, "Introduction to the Single Text Relating to the Mandate of the First Committee," p. 6.

12. Miles, op. cit.

13. Ibid. and E. Miles, "The Dynamics of Global Ocean Politics," in Douglas Johnston (ed.), Marine Policy and the Coastal Community: Studies in the Social Sciences (London: Croom Helm, 1976).

14. A/CONF.62/37, and A/CONF.62/41.

15. A/CONF.62/C.3/SR.22.

16. Miles in Johnston, op. cit. identifies the members of the territorialist group as Brazil, Ecuador, El Salvador, Nicaragua, Panama, Peru, Uruguay, Dahomey, Guinea, and Somalia.

17. B. G. Buzan and B. Johnson, Canada at the Third Law of the Sea Conference: Policy, Role and Prospects, Law of the Sea Institute Occasional Paper, no. 29 (Kingston: University of Rhode Island, 1975).

18. A/CONF.62/SR.56, p. 4.

19. Ibid.

20. Alternatives to a single treaty are discussed by Douglas M. Johnston, "The Options for LOS III: Appraisal and Proposal," in Christy et al., op. cit., pp. 357-72.

CHAPTER

11

SUMMARY AND CONCLUSIONS

In the preceding ten chapters we have been examining chronologically the change from a situation in which there was very little interest in most of the seabed to a situation in which the entire seabed is, or is about to be, covered by a variety of explicit legal regimes.

Since the limits of that approach have been reached, it is worthwhile now to reconsider the data in the light of the three aspects of the political process outlined in the introduction, namely economic and technological factors tending to raise the value of the seabed; actions of states, particularly claims, in relation to the seabed; and actions within international organizations in relation to the seabed. The intention in this chapter is to link these three aspects together as the nodes of a simple model of seabed politics, and to review them in the light of the three basic questions also posed in the introduction: why and how did the seabed become an international problem; how was the problem solved in terms of a given range of options; and what factors were responsible for the nature of that solution? Particular attention will be paid to the last of these questions in relation to the more general question of whether or not the nature of the international negotiating process precludes the radical developments in international organization necessary to implement a solution involving international jurisdiction and supervision over valuable economic resources.

ECONOMIC AND TECHNICAL FACTORS

The first question—Why and how did the seabed become an international problem?—is largely answered by an examination of the first node—economic and technical factors tending to raise the value of the seabed.

Before 1850, the seabed was a little known area of almost no value. Lack of appropriate technology made all but a small fraction of it inaccessible, and such claims to it as there were related either to shallow areas rich in sedentary living resources, or else to the simple association of seabed with territorial sea claims. Three linked factors changed this situation: the development of technology applicable to the seabed, the changing pattern of use of the waters overlying the continental shelf, and the worldwide increase in demand for almost all types of living and mineral resources.

The development of technology applicable to the seabed started early and went through a long period during which it did much to expand knowledge about, and usefulness of, the seabed, but little to increase its economic value. This included lead and line depth surveys for navigational and scientific purposes, all the late 19th century achievements in cable-laying, and many of the advances made in submarine and anti-submarine warfare since the beginning of the present century. It was not until the end of the 19th century that developments in drilling from platforms and mining by dredge began to raise the economic value of the seabed. For several decades, these developments continued at a slow pace, and so did not raise the value of the seabed outside the narrow bounds of existing territorial seas.

Only after World War II did the capabilities of seabed mining technology begin to expand rapidly and influence the value of the seabed beyond the territorial sea. The breakthrough came in the oil industry with the invention of the floating drill rig in 1947. Thereafter, the offshore industry expanded rapidly, first into select shelf areas like the Persian Gulf, Venezuela, and California, and then during the 1960s and 1970s into nearly all the continental shelf areas of the globe. Extraordinary improvements in technology enabled the offshore oil industry to turn virtually the entire continental margin of the planet into highly valued real estate (either in fact, or by speculative association) in the space of three decades.

The oil industry, however, did not much affect the value of the deep seabed beyond the margin. The value of this area did not begin to rise until the mid-1960s under the impact of commercial interest in manganese nodules. With the rapid development of technology for nodule mining starting in the late 1960s, the hard mineral industry leapfrogged over the oil industry, to which it had always been smaller brother on the continental shelf, and made nearly the entire seabed accessible to commercial interests. In the case of both the oil and hard mineral industries, the nonrenewable character of the resources which, in conjunction with the technology, gave the seabed its value played a vital role in undermining the old freedom of the seas regime, which was heavily based on the assumption that ocean resources were inexhaustible.

The older scientific and military interests in the seabed expanded in step with commercial interests and added their substantial weight to the massive developments of technology relevant to the seabed. From the 1960s onward, the United States and the Soviet Union were especially active in this way.

The economic value of the seabed, in particular the continental shelf, was also raised by its association with the value of fishing grounds in the overlying waters. Nearly all the world's major fishing grounds occur over the shelf, and by the 1930s signs of conflict between maritime and coastal states over these resources were becoming manifest. The rapid increase in the intensity of fishing after World War II brought into question the assumption of inexhaustibility even of living resources, and precipitated the conflict between maritime and coastal states for control of these resources, most of which lay beyond the traditional three-mile territorial sea. The value of the seabed here was that the geographical relationship between the continental shelf and the coastal state could be, and was, used to support claims by coastal states to extended jurisdiction and sovereignty over the superjacent waters and their living resources.

The steady increase in demand for resources of all kinds naturally underlay both the technological and fishing reasons for the rise in the value of the seabed. The primary factor here was population growth linked with rising expectations of improved standards of living all over the globe. The desire to gain access to secure supplies of resources that had run short within its land territory was the major motive behind the leading role played by the United States in the development of both offshore oil and manganese nodule mining technology. In this regard artificial shortages, such as that created by the OPEC states during 1973-74 in oil and subsequently threatened by other commodity-producing organizations, have just as great a stimulating effect on the value of alternative resources, and therefore on the development of seabed technology, as real shortages.

The rise in the economic value of the seabed caused by these three factors was accompanied by increased demands to exercise control or ownership over the seabed and its resources. The rise in value of the continental shelf following World War II motivated the series of claims, starting with the Truman Proclamation in 1945, which elevated the seabed problem to the international level in the work of the ILC and the First UN Conference on the Law of the Sea in 1958. Likewise, the rise in value of the deep seabed with the prospect of nodule mining in the mid-1960s precipitated the international response starting in 1967 in the SBC and continuing through the mid-1970s at the Third UN Conference on the Law of the Sea. In both cases, it should be noted, the demand for control arose well before any significant wealth was being produced from the areas in question. This indicates the sensitivity of the political process to increases in economic value, a sensitivity that could have dramatic implications for other "unowned" areas, for example, Antarctica.

The rise in the economic value of the seabed, then, largely explains why and how the seabed became politicized. Furthermore, the way in which the seabed became an international problem suggests much about the relationship between a rise in economic value and the other two nodes of the model. The rise in value of the seabed directly stimulated state action in the form of unilateral claims to the continental shelf during the 1940s and 1950s. These claims, in turn, stimulated international action in the form of the ILC and the UN conferences. A rise in value can, under unusual circumstances, also stimulate international action directly, as in the case of the UN response to the prospect of nodule mining on the deep seabed in the mid-1960s. This raises the problem, to be explored later, of a conflict developing between state and international organization action if both respond independently to a rise in value. In addition, a feedback effect should be noted from both state and international levels, in that provision of a stable legal regime for an economically valuable area enhances its value by encouraging exploitation of the resources. International action can also, of course, influence state action, as in the case of the post-1958 state claims to the continental shelf. In simple form, the model can be represented as in Figure 11.1.

STATE ACTION

Examination of the second node of the model—state action—adds detail to the answer to the question of how the seabed became an international problem. More important, however, it provides a substantial part of the answer to how the problem of creating a legal regime for the seabed was solved, and what forces shaped the character of the solution.

The victory in the first round of creating a legal regime for the seabed clearly went to the coastal states. The numerous unilateral state claims to the continental shelf that followed the Truman Proclamation met no opposition in terms of coastal state rights to exclusive jurisdiction over the mineral resources of the shelf. Where controversy did arise was over the living resources in the superjacent waters, and for this reason the issue of coastal state

FIGURE 11.1

An Interaction Model for Seabed Politics

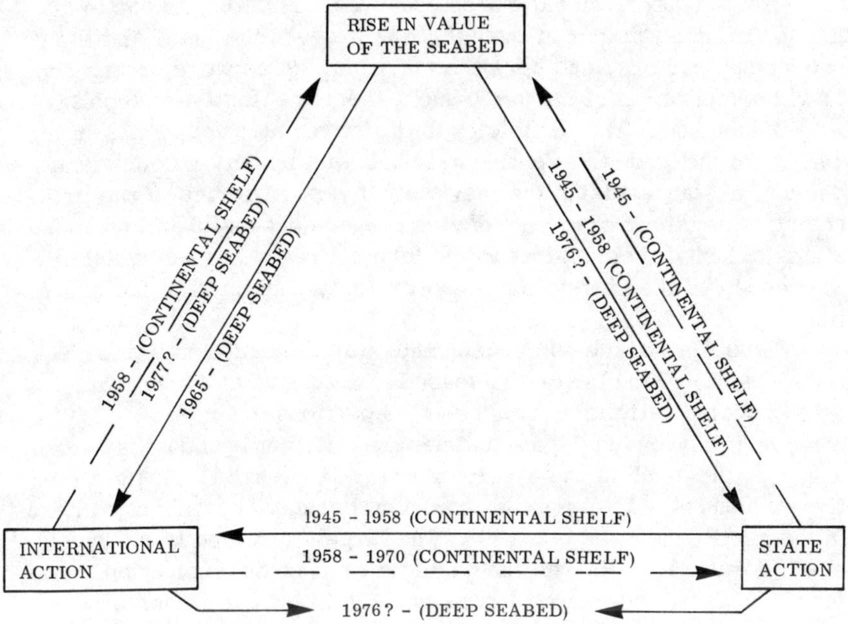

Legend: Prime Motive ———→
Feedback Effect — — — →
Speculation ?

Note: In terms of the model, state action is interpreted as claims to jurisdiction or sovereignty over the seabed. The possibility of a link between state action in this sense and state action relating to the economics of a rise in value of the seabed is a difficulty in the model. An example of this might be the U.S. government's sponsorship of deep water technology starting in the early 1960s and possible links between this and the development of nodule mining technology by U.S. corporations during the later 1960s. Should the United States pass its own nodule mining legislation, the link between the legislation and the American corporations could call into question the blindness of the forces underlying the rise in value of the deep seabed. This would complicate, but not invalidate, the model.

Source: Compiled by the author.

sovereignty over the shelf itself was extremely sensitive. The 1958 UN conference simply confirmed the coastal state initiative in claiming exclusive control over continental shelf resources. No attempt was made to leave the shelf under the high seas rules, and only a tiny effort at internationalization occurred. Most of the battle at the conference was fought over the link between the shelf and the superjacent waters, and here the maritime states were strong enough to win a separation of the two issues, thereby effectively depoliticizing the seabed. The flexible exploitation formula for limits in the 1958 Convention on the Continental Shelf was merely a convenient device for standardizing the inevitably divergent forms of the preceding national claims. The device proved successful in that most states adjusted their claims to the formula regardless of whether they accepted the restriction on sovereignty contained in the convention.

The upshot of the 1958 conference was the confirmation of coastal state control over the seabed in a zone of indeterminate width covering all the most accessible portions of the seabed. The zone certainly included the continental shelf, but boundaries became vague as the depth of the overlying waters increased. At the time this vagueness did not appear important because the primary interest in the shelf concerned oil, which was largely confined to relatively shallow waters. Although the absolute limit of the shelf, and therefore of coastal state control, was unclear in law, in terms of what interpretations of the convention states would be likely to accept it probably included the margin but not the deep seabed. This view was supported by the ICJ in its influential 1969 decision on the North Sea shelf.

After 1958, state action on the seabed was much less significant than it had been before. With coastal state rights to the valuable parts of the seabed secure, the only coastal state action of importance to the seabed issue were claims to extended territorial seas and maritime zones. These claims related mostly to the ongoing dispute over fishing rights, which had not been settled at the first and second UN conferences. The fact that they included seabed areas beyond the edge of the margin was coincidental, for at the time these areas of seabed were not considered to be valuable.

By 1970, the relatively minor matter of the flexible limit was under attack for two reasons: first, the need to establish a fixed boundary for the international seabed area; and second, the growing weight of fixed limit territorial sea and maritime zone claims. It is important to note in this context that the essential concept of coastal state control over the seabed zone adjacent to the coast was not in question, except in a limited way from the landlocked states. The fairly rapid evolution of the 200-mile/margin fixed limit for coastal state control left the real substance of the 1958 agreement on the shelf relatively untouched in terms of seabed matters.

The shaping force that determined this part of the seabed regime was unquestionably the coastal state movement. This movement has been described in preceding chapters as the expression of a historical imperative relating not just to seabed matters but to the whole spectrum of maritime jurisdiction. The movement resulted from a number of factors, nearly all of which tended to attract developing countries as the backbone of its membership. Since the underlying cause of the rise in coastal state awareness was the exhaustibility of ocean resources, both living and mineral, those coastal states not heavily engaged in exploitation of marine resources were the first to realize that they had something to lose. So long as ocean resources were considered to be inexhaustible, the exploiting activities of maritime states were not prejudicial to anyone else's interests. As soon as that ceased to be the case, however, the exploiting activities of maritime states became a direct loss to the coastal state off whose shores it occurred. Since nearly all the major coastal states among the developed countries were, on the basis of their technological and economic strength, maritime states, the burden of coastal state awareness fell onto the nonmaritime coastal states among the developing countries.

Because of this link, the coastal state movement became largely, though by no means fully, associated with what might be called the developing countries movement. The acute sensitivity of the developing countries, many of them newly independent, to matters of nationalism, territorialism, and control over resources relating to economic development blended easily with the demand for control over ocean territory and resources that was the essence of the coastal state movement. In addition, both movements were characterized by strong distrust of the developed states, particularly in regard to economic exploitation based on superior technology. Given this association between the two movements, the coastal state movement naturally gained strength as the decolonization process following World War II swelled the ranks of the developing countries. This was particularly true during the 1960s with the massive influx of African and other new states (none of which had participated in the making of the existing law of the sea) into the international system, and with the founding of organizations like the Group of 77 to give expression to the developing countries' point of view.

The Latin American states unquestionably played a leading role in the coastal state movement. This was partly a natural development, based on the fact that the Latin American states were the oldest developing countries, and partly a geographical development, based on the concentration of states with large coastal interests in Latin

America. It was in Latin America that the combination of national territorialism, technological inferiority, rising demand for resources, lack of maritime traditions and interests, and exploitation off the coast by foreign maritime powers first generated coastal state awareness and activity. Coastal state activity became apparent in the 1930s with the concentration of territorial sea claims greater than three miles. It became intense after World War II with the Latin American regional activity of the 1950s, and with the flood of extensive claims to the continental shelf, epicontinental sea, territorial sea, and maritime zone that continued right up into the 1970s. One of the results of the regional activity, the Principles of Mexico, can be seen as the first manifesto of the coastal state movement. During the 1970s, the Latin American coastal states led the struggle for the 200-mile limit, which was based on Latin American unilateral precedents and eventually carried most other coastal states with them on the question of limits, and many, though not all, on the more divided question of rights.

Latin American leadership of the coastal state movement was also facilitated by three other factors. First, the prevailing circumstances in the international system were such as to encourage the process of unilateral claims. Not only did international law provide an incentive for such claims in the form of allowing customary law to develop by state practice, but claims themselves tended to spread much faster than the particular interests of states would ordinarily dictate. Only a few states were directly motivated in their claims by economic circumstances, for example the United States by continental shelf oil, and Chile, Ecuador, and Peru by geographical circumstances and fishing problems. The actions of these few, however, precipitated a rush of similar claims from states with much less obvious interests in the issue. Some of these were made for underlying reasons (for example, claiming the shelf in order to claim overlying fisheries), some simply as bandwagon responses or simple territorial aggrandizement, and some, eventually, as a response to the standardizing drive of international law as created through international organizations.

Second, Latin American leadership was facilitated by the general amenability of the developing countries movement to leadership from states, or groups of states, with a policy in line with the movement's overall aim of redressing the economic and political balance between developed and developing countries. While this amenability was by no means uncritical (as illustrated by the more moderate coastal state position evolved by the African states), it did provide a sympathetic and organized forum in which developing states could seek support for policy initiatives. Other examples of this phenomenon (often dismissed in developed states as a process of the unscrupulous leading the ignorant) were the Group of 77's support for the OPEC actions on oil and for the demands of the metal-exporting states for protection from competition by seabed mining. The anti-developed state bonds of the developing countries movement were often strong enough to contain policies deleterious to the immediate interests of a majority of developing countries but in line with the movement's long-term objectives.

The third factor facilitating Latin American leadership, and indeed the development of the seabed aspect of the coastal state movement as a whole, was the lack of any countervailing force. This was especially true on the seabed, but much less so for the water column, where the maritime states put up strong resistance. The opposition of the landlocked and geographically disadvantaged states to coastal state expansion on the continental shelf came too late, and was too weak, to have much effect.

The great strength of this coastal state historical imperative was further indicated by the conversion of developed states like Canada, Australia, New Zealand, Iceland, and Norway into some of its leading exponents. These states all downplayed their European maritime traditions, which were mostly not necessitated by possession of large maritime interests, in favor of coastal geographic and economic interests. In addition, significant inroads have been made into the ranks of the major maritime powers, with countries like Britain and the United States giving higher priority to the protection of some coastal interests. The coastal state movement thus unquestionably dominated the first stage of creation of a legal regime for the seabed, during which nearly all the shallower, more accessible areas of the seabed adjacent to land masses came under coastal state control.

ACTIONS WITHIN INTERNATIONAL ORGANIZATIONS

The third node of the model—action within international organizations—is at once the most difficult and the most rewarding to examine. While it adds only a little to the account of why and how the seabed became an international issue, it provides extensive information relevant to the other two basic questions under consideration here. In terms of outcomes, it includes the 1958 Geneva Conventions and the numerous UN resolutions on the subject since 1967. Unfortunately a final outcome for the Third UN Conference on the Law of the Sea is still some years in the future, but even at this date the UN documents provide very clear outlines of the likely new legal regime. In terms of the forces shaping the solution, however, this is the richest node of the three. By its very nature, action within international organization results in the visible politicization of an issue on a global scale, and reveals in great detail the interplay of forces engaged.

In addition to its value in revealing shaping forces, action within international organizations raises interesting questions about the power of taking the initiative in seabed politics, and indeed in international politics generally. Having the initiative in international politics is rather like possessing "the tempo" in chess or "sente" in go—it often gives one player a commanding advantage in a situation of otherwise equal power and position. Usually the initiative falls to states, in part because of their primacy in the international system as sovereign entities and in part because international organizations are slow to move because of the organizational and political restraints on their decision-making powers. As we have seen, however, seabed politics offers two contrasting case studies on the location of initiative: the first from 1945 to 1958, during which time initiative lay with the coastal states and the actions of international organization were largely in response to the actions of such states; and the second from 1967 to the present, during which time initiative lay within the United Nations and the whole framework of seabed politics (relating to the deep seabed) was formed within the UN and not by the claims of states.

In the first case, states responded directly to a rise in value of the seabed by making claims to the continental shelf. International organization responded to state actions by standardizing and depoliticizing the claims. In the second case, the UN responded to a rise in value of the deep seabed before any states had made claims. Two factors in particular account for the UN's unusual seizure of the initiative in this case: the relative physical remoteness of the deep seabed in contrast to the continental margin, and the unique idiosyncratic role played by Pardo in using the UN as a platform from which to launch his carefully aimed proposal.

Given that the initiative in the second phase of seabed politics lies within international organization, the question that arises is: "What happens?" Obviously the process leading from the seizure of the initiative (Pardo's proposal) to a policy outcome (a legal regime for the deep seabed) is much more complicated within an international organization than it is within a state. If this process is too cumbersome to work either because of its basic nature or because it can be made unworkable by political maneuvering, then the prospects for future initiative-taking within the United Nations aimed at radical reform of the international system are rather dismal. The seabed issue has special significance in this regard because it involves creation of a potentially unprecedented type of international organization. Since expansion of the structures, powers, and functions of international organization as an alternative, rather than as a complement, to an international system dominated by powerful states seems likely to occur only through initiative at the international level, the incapacity of the present international organization to sustain a strong and promising initiative in that direction would be a blow to the hopes of those who place a high value on such development.

This brings us back to the question of how the outcome of seabed politics is shaped, and here analysis will focus on the influence of a wide range of factors, including the structural characteristics of the negotiations in terms of actors, alignments and procedures; the nature of the issues in terms of number, complexity, and linkage; the powers that contending actors can bring to bear in support of their position; and external factors such as historical precedent, activity on separate but related issues, economic and technological developments, and so forth. These factors will be examined for two types of influence on action at the international level: first, factors that shape the process of action in terms of speeding it up or slowing it down, and weakening or sustaining the commitment of states to action at that level; and second, factors that shape the substance of action, including the important interaction between the nature of the process and the nature of the substance.

In the pages following, this framework will be used to examine the international negotiations on the seabed from 1968 to 1975. The sequence will be first to review the factors influencing the process of the negotiations as a whole, aside from seabed issues, and then to look at the factors shaping process and substance on the continental shelf and deep seabed issues in particular. This exercise has four purposes: first, to identify the shaping factors and their effects; second, to identify the influence of substance on process and process on substance; third, to comment on likely trends in the negotiations; and fourth, to marshall evidence on the question of whether or not the process leading from initiative to policy outcome at the international level is workable if the initiative involves radical advances in the structures, powers, and functions of international organization.

FACTORS SHAPING THE PROCESS OF THE NEGOTIATIONS AS A WHOLE

The factors applicable to the post-1967 law of the sea negotiations in general all shaped process, and only influenced substance inasmuch as process factors in general bore on particular substantive issues.

Speed

In terms of the speed with which the negotiations were conducted, the balance tipped rather heavily in favor of those factors tending to slow the process down. These were mostly concrete in nature, affecting directly the behavior of delegates.[1] They included the large number of states involved in the negotiation, each with the right to express its views on every issue; the ignorance of many delegations about the technical and economic aspects of law of the sea issues; the large number of issues involved in the negotiation, many of which were formidably complicated in their own right and many of which were complicated by interaction with other issues both within and outside the negotiations; the consensus procedure, with its built-in delaying mechanisms; the proliferation of policy alternatives; and the general difficulty of legal drafting at the international level, particularly in the absence of an initial single draft text. In addition, the lack of firm deadlines affected the attitude of delegates in such a way as to slow down the negotiations. The only effective deadline in this regard was the one that precipitated the flurry of activity surrounding the declaration of principles in 1970. All of these slowing factors would have operated even if the commitment of states to the negotiations had been uniformly high. The fact that some states—most notoriously the Latin American territorialists—had a very low commitment greatly exacerbated the problem by opening the whole process to deliberate obstructionism.

By contrast, the main factors tending to speed things up were mostly abstract in nature, affecting the attitudes of delegates rather than shaping their behavior directly. These included the cost of lengthy negotiations in terms of both funds and allocation of scarce diplomatic resources; the prospect of unilateralism, and the associated prospect of conflict, arising out of failure to conclude an agreed law of the sea convention; and the pressing need for solutions to a whole range of existing maritime problems, ranging from fisheries through navigation to nodule mining, in all of which legal uncertainty made for bad economics. All of these factors, sometimes reinforced by domestic political activity, tended to impart a sense of urgency to the earliest possible conclusion of the negotiations. In addition, the elimination of peripheral alternatives from the agenda speeded up the negotiations, but neither the SBC nor the Third Law of the Sea Conference had a distinguished record in this regard.

Commitment

In terms of the commitment of states to action at the international level, the balance of factors affecting the law of the sea negotiations as a whole tipped definitely, but not heavily, in favor of those tending to sustain such commitment. These included the personal prestige of the delegates and officials involved in the negotiations; the momentum created by the enormous volume of work already done; the perception of interests common to nearly all states that could best be served at the international level, for example, avoidance of military conflict, standardization of international law, and the maintenance of a free-flowing physical structure for world trade by sea; the influence of international organization in support of the negotiations; and satisfaction with, or at least acceptance of, the substantive trend of the negotiations by most of the states possessing strong unilateral options on law of the sea matters.

Those factors tending to weaken commitment included unilateral claims by states; frustration with the slowness of the negotiations, particularly on the part of states with urgent maritime problems; and frustration with the substantive direction of the negotiations, particularly on the part of maritime, landlocked, and geographically disadvantaged states. While these factors were the subject of much comment, both by delegates and observers, the commitment of states to the ever extending negotiations has in fact remained remarkably high, as demonstrated by their exercise of restraint in terms of unilateralism and by their continued participation in the negotiations despite the violation of deadlines.

In summary, then, the forces shaping the process of the law of the sea negotiations in general tended to make those negotiations stable over the short and medium term in the sense that states were fairly solidly committed to them, but slow and therefore excessively drawn out. These two trends made for a potentially unstable situation in the long run, because slowness was one of the factors undermining commitment both directly and indirectly in that it provided time and incentive for unilateral action by states. The general affliction of slowness had powerful but quite different effects on the two seabed issues under consideration here.

FACTORS SHAPING THE NEGOTIATIONS ON THE CONTINENTAL SHELF

During the second round of seabed politics, starting in 1967, the continental shelf was not a live issue in the same sense as the deep seabed. The basic allocation of the shelf to the coastal states was complete by 1958, and what was at issue during the second round was much more the question of a fixed limit for coastal state jurisdiction than the question of coastal state rights per se. On the limits question, initiative still lay with the coastal states and not, as for the deep seabed question, with international organization.

Speed

In terms of the speed of the negotiations, the balance of factors on the continental shelf issue tended to add to the burden of slowness already imposed by general factors. The main factor facilitating these negotiations was the vigorous leadership from the Latin American states on the 200-mile limit. Especially after 1970, this leadership provided a solid focus for the discussions by introducing a fixed alternative to the by then unacceptable flexible limit for the shelf. Other factors also helped to move the negotiations along, but none of them provided a major counterweight to the numerous forces working in the opposite direction. These minor factors included the declining emphasis on the work of C.I after 1969, the fear of delimitation conflicts relating to continental shelf claims, and the virtual elimination of the narrow limits alternative after 1973.

Against these, however, stood an imposing array of factors tending to slow things down. These included the conceptual and political dilemma over whether to negotiate the boundary or the deep seabed regime first, which paralyzed the debate until at least 1970; the steady rise in the value of the shelf, which intensified various boundary and rights disputes; the well-organized and concerted opposition of the landlocked and geographically disadvantaged states first to wide limits and later to exclusive coastal state rights (this opposition was in part a reaction to the Latin American leadership, and therefore cancelled much of its accelerating effect); and the linking of the continental shelf issue to the whole economic zone question after 1970, a link that enormously complicated solution of the relatively straightforward seabed question.

The major effect of slowness on the continental shelf negotiations was to strengthen the hand of the coastal states. These states were in a strong position anyway, not only because of the legitimacy given to their unilateral claims by the conventions of customary law and because of their option to make such claims if they had not already done so, but also because of the weight of existing international law giving them rights to the shelf in the 1958 convention and the 1969 ICJ judgment. The passage of time added to their advantages by giving them opportunities to entrench their claims, to encourage others to make claims, and to wring concessions from opponents who mostly held much weaker bargaining positions. The activities of the Latin American coastal states, particularly the territorialists, illustrate that the advantages of drawn-out negotiations were not lost upon the delegations of the coastal states. In this sense also, the campaign of the landlocked and geographically disadvantaged states was to a considerable extent self-defeating: inasmuch as it lengthened the negotiations, it strengthened the position of the most intransigent coastal states.

Commitment

In terms of states' commitment to the shelf negotiations the alignment of factors supported maintenance of the international approach quite strongly. The only real weakening factor was the unilateralism of the territorialist states, and this applied much more to water column than to seabed issues. Unilateralism in the form of claims to the shelf itself was not an important problem in the second phase of seabed politics. The main sustaining factor was the need to establish an internationally agreed fixed boundary for the international seabed area, a task that would be exceedingly difficult outside the international framework, and that was very widely supported as necessary. In addition, the large body of landlocked and geographically disadvantaged states carried a strong commitment to the international level of action on this issue, because it was the only level other than the regional that offered them any possibility of influencing the outcome. There was little incentive, then, for any group of states to abandon the international approach on the question of continental shelf limits so long as the creation of a deep seabed regime remained a generally supported objective. This lowered the probability of states preferring unilateral or regional solutions on the margin question, though it did not affect the use of the threat of such solutions as a bargaining lever in the international negotiations.

Substance

Any assessment of the factors shaping the substance of the continental shelf negotiations must necessarily be tentative at this time because of the absence of a final document from the Law of the Sea Conference. Nevertheless, a fairly conclusive assessment is possible because, as we have seen, the single negotiating text arising out of the Geneva session contained a set of compromises that looked quite final in terms of the relative strengths of the alignments behind each of the remaining policy alternatives.

The key background shaping factor was the political and legal framework of the 1958 convention and the 1969 ICJ judgment. This pre-existing framework firmly established the principle of coastal state jurisdiction over at least the major part of the continental margin, and thereby made exceedingly difficult the introduction of any scheme for substantial rollbacks of coastal state jurisdiction in favor of the international seabed regime. It helped put the coastal states in a strong position not only because it gave them prior rights generally, but also because the dramatic advances in offshore oil technology and activity enabled them to take maximum advantage of the flexible exploitation criterion for determining the outer limits of their jurisdiction.

Given this basic framework, two factors combined to define the central question pertaining to the continental shelf issue during the second round of seabed politics. The question was, "What fixed limit should be placed on coastal state seabed jurisdiction?" The two factors were Pardo's proposal for an international regime on the deep seabed and the 200-mile territorial sea and maritime zone claims of the Latin American states. Both these factors undermined the flexible limit compromise, and both were themselves stimulated and aided by advances in seabed oil and nodule mining technology. In their different ways, these two factors illustrate the power to shape events that is inherent in a well-conceived seizure of the initiative.

Finally, given both the basic framework and the definition of the central question, the last set of shaping factors was the various groups and schools that formulated and contested the alternative solutions to the central question. In the case of the continental shelf, the struggle took place between a strongly organized but fundamentally weak group, the landlocked and geographically disadvantaged states, and a very weakly organized but fundamentally strong collection of coastal states. In the emergent solution on this issue the landlocked and geographically disadvantaged states have fallen far short of their stated objectives, having achieved only the coastal fisheries trade off and the compromise on sharing from the margin beyond 200 miles. This occurred despite both the group's efficient organization and its strong influence in the African group and the flat 200-mile limit school, and can be explained by its almost complete lack of power outside the format of an international conference. Had the group been able to establish closer ties to the United States, which for a time occupied a middle ground position, it might have exercised much more influence on the outcome, but after 1970 the United States drifted toward the coastal state position for largely domestic reasons.

The coastal states, by contrast, had almost everything working in their favor despite being divided among themselves on policy. The power given to the coastal states by factors ranging from the historical imperative, through the slowness of the international negotiating process, to the immense leverage of the unilateral claim or option, fairly easily overcame the difficulties of conflicting schools (200 miles versus margin), conflicting groups (moderate coastal versus territorialist), and entanglement with controversial water column issues (coastal versus maritime interests). The end result looked rather like a repeat of the 1958 settlement in that the international negotiating process was being used to legitimize and standardize claims made by states (200 miles and edge of margin) on their own initiative. In this regard, the standardizing drive of international law-making gave enormous influence to the Latin American "vanguard" states by enabling them to enshrine the limits aspect of their maximalist unilateral claims as the basis for an international standard. Seizure and skilled use of the initiative by these states paid off handsomely in terms of seabed politics, and substantially, if at the moment less clearly, in terms of the more complex and controversial politics surrounding other aspects of the economic zone.

FACTORS SHAPING THE NEGOTIATIONS ON THE INTERNATIONAL SEABED REGIME AND MACHINERY

The question of what factors shaped the outcome on the issue of the international seabed regime and machinery is both more difficult and more significant than the same question on the continental shelf. The major difficulty is that the final outcome on the international regime and machinery is not yet either as clearly defined or as likely to be implemented as is the case with the outcome on the continental shelf. Powerful alignments supporting opposed policies still exist, and it is not even wholly certain that the outcome will be some form of supervised international jurisdiction. Despite this, sufficient progress has been made in the negotiations to allow for a substantial analysis of shaping factors. Indeed, with the passing of the 1975 deadline, the negotiations have reached a turning point at which such an analysis is not only possible, but necessary to an understanding of what is happening. The significance of the question lies in its close relationship to initiative-taking by international organization, and in particular to the process leading from initiative to policy outcome on the issue of creating a new type of international organization.

FIGURE 11.2

Summary of Key Factors Shaping Negotiations on the Continental Shelf

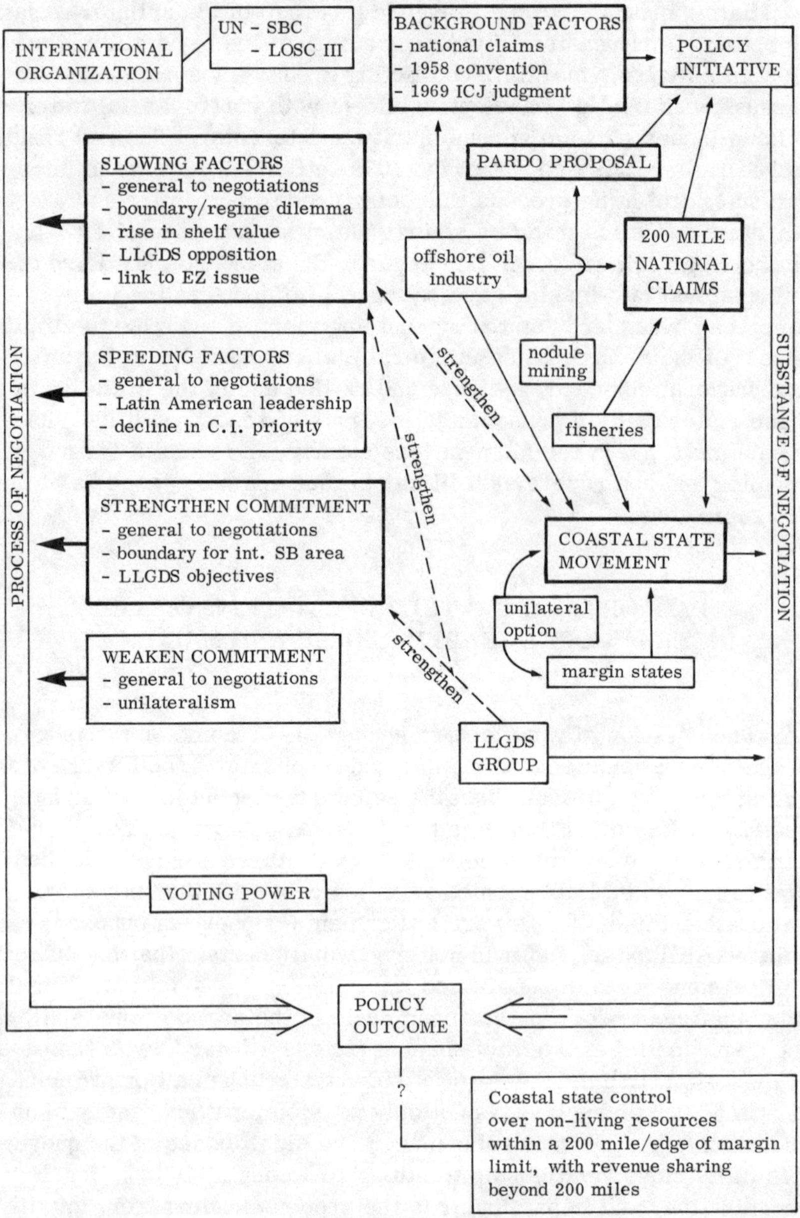

Note: The dominant set of shaping factors are set in bold squares.
Source: Compiled by the author.

Speed

In terms of the speed of the negotiations, the balance of factors on the international seabed regime and machinery added heavily to the overall weight of factors tending to slow down the negotiating process at the Law of the Sea Conference.

Very few factors can be identified as acting to speed up the process in any noticeable way. The prospect of unilateral action by the advanced technology states has often been cited as a stimulus to the negotiations, but in practice does not appear to have had this effect. The Metcalf bill, far from cowing the developing countries into making concessions, contributed to their unity by emphasizing the ideological aspects of the issue. Their reaction against the bill has, if anything, prompted them to stake their position on a point of principle at the risk of losing all to unilateralism by the advanced technology states. In view of this, the only factor facilitating negotiations on this issue has been the early elimination of many theoretically possible regime models from the debate resulting from the rapid polarization of states along developed/developing lines.

By contrast, the list of factors tending to slow down negotiations on this issue is long and formidable. First there were the conflicts built into Pardo's proposal, which had such a great influence on the subsequent negotiations. These included especially the developed/developing conflict format in which Pardo cast his whole proposal, and the hopelessly confused and controversial "peaceful purposes" provisions he chose to stress. Second was the vague mandate given to the AHSBC, which, along with the dearth of appropriate information, made it difficult for the committee to move decisively in any one direction. Third was the decline in the perceived value of the deep seabed after 1969, which followed the optimistic assessments surrounding the early interest in nodules. This not only lowered interest in the issue, but also facilitated a shift in the discussion to a more abstract and intractable political level. Fourth, and related to this, was the ongoing, oil-based rise in the value of continental margin areas, which drew interest away from the deep seabed regime and encouraged a general shift of emphasis in the law of the sea negotiations toward coastal zone matters.

Fifth was the nature of the political balance between the two opposing sides on this issue. The key point here is that although each side commanded wholly different forms of power and influence in relation to the issue, they were so evenly matched that neither could decisively overwhelm or outmaneuver the other. The Group of 77, on the one hand, possessed not only a huge voting majority in the relevant international organizations, but also the power of its members to take unilateral action in relation to other law of the sea issues of importance to the advanced technology states. In addition, since Pardo had essentially designed his proposal as a policy for the Group of 77, the Group benefited from the tremendous shaping influences that those proposals had on the negotiations.

The advanced technology states, on the other hand, controlled all nodule mining technology, which gave them not only a unilateral option on the matter but also the power to render useless any international machinery set up against their will simply by refusing to cooperate with it. In addition, of course, they also possessed the traditional economic and military advantages of the great powers. This situation meant that the only possible solution for both sides at the international level was one on which they could genuinely agree. The process of negotiating such a genuine agreement was made exceptionally slow not only by all the factors slowing down the law of the sea negotiations generally, but also by the depth of disagreement between the two sides, by the ideological content of the debate, by the "time is on our side" mentality that affected both sides for different reasons (technological options for the advanced technology states, and a general political euphoria on the part of the Group of 77), and by the enormous complexity of the issue as a whole.

The complexity of the deep seabed regime and machinery issue ranks as a separate, sixth, factor tending to slow down the negotiations. Particular problems arising here included the dilemma over whether to define the area or the regime first, the economic and technological unknowns of nodule mining which contributed to the politicization of information by both the advanced technology states and the mineral exporting states, the location of power and initiative in the structure of the seabed authority, and design of compensatory mechanisms for states adversely affected by exploitation of the common heritage. In addition, there were problems of interaction between deep seabed and other issues, including the regime of islands, high seas fisheries, scientific research, the global commodity debate, and the transfer of technology.

A seventh factor loomed into significance during the Geneva session, and that was the possibility that substantial numbers of key states were losing their commitment to international negotiations on this issue.

Commitment

In terms of states' commitment to the deep seabed regime negotiations, the balance of forces appeared to fall slightly on the side of sustaining such commitment, and was sensitive to a number of variables, particularly the passage of time.

Three factors tended to weaken commitment. First was the coastal state movement which, especially after 1969, drew attention away from the deep seabed issue. This factor, however, was more significant in slowing rather than actually threatening the deep seabed negotiations. Second was the growth among some of the developing countries of sentiment against any exploitation from the deep seabed. This arose partly among states with competing mineral-exporting interests, and partly among states with ideological interests that accepted the argument that under any circumstances the industrialized states would be the primary beneficiaries of nodule mining. The threat posed by this factor varied with the strength of the influence wielded by these states in the Group of 77. Up to the end of the Geneva session, they had proven an active but not dominant influence within the Group, and the possibility that the global commodity debate would strengthen their hand remained an active threat to the deep seabed negotiations. Third was the possibility that the United States would enact some form of ocean mining legislation or that the mining companies would begin commercial operations without such legislation. Such action would be very likely to enrage the Group of 77, and consequently to jeopardize the whole Law of the Sea Conference.

The slowness of the negotiations influenced all three of these factors. By the summer of 1975, it was tending to reduce the influence of the coastal state movement because coastal issues were moving toward resolution on some fronts. This was not true of either anti-exploitation sentiment or possibilities of unilateral action, where the passage of time tended to increase the likelihood of damage to the international negotiations. The fact that these two problems also exacerbated each other created a potentially explosive situation that was one of the major threats to the continuation of the conference. Such a situation was, however, only a potential, and up to the autumn of 1975 had been kept from becoming reality by the countervailing factors tending to maintain states' commitment to the international negotiations.

The factors tending to sustain commitment related mostly to the fundamentally international character of the deep seabed issue. This character stemmed originally from the freedom of the seas regime traditionally applicable to the area, and was greatly reinforced by Pardo's proposal and the resulting declaration of principles passed by the General Assembly in 1970, both of which added momentum to the movement toward an international solution. Any movement toward unilateral solutions thus went against a powerful mainstream of traditions, opinions, and events. In so doing, it raised possibilities of conflict, and also of disruption of common interests on other law of the sea matters, that no major party to the law of the sea negotiations desired.

The Group of 77 in particular had a strong commitment to the international negotiations on this issue. First, since its members had no really credible unilateral options in relation to the deep seabed, an international solution offered them the best, and possibly the only, chance to influence the issue. Second, the Group had a general commitment to strengthening international organization as a means of redressing the economic and political balance between developed and developing countries. Pardo aimed his proposal accurately at this interest, and the result was the enshrinement of the international seabed regime and machinery as a test case in the Group's overall campaign. The subsequent rise of the global commodity debate and the new economic order movement acted to strengthen its commitment, though there was danger here that the deep seabed issue would be wholly absorbed by this larger debate and consequently become even further politicized and deferred. In addition, the Group's commitment was reinforced by internal politics relating to the landlocked and geographically disadvantaged states, which formed a large contingent within the Group. Both the politics of internal unity and the Group's general sense of egalitarianism demanded that these states receive something to balance the gains made by coastal states in the large extensions of maritime jurisdiction. The deep seabed regime served this purpose, and was therefore an important part of the policy package necessary to maintain Group unity.

The commitment of the developed states to an international solution on the deep seabed appeared less strong than that of the developing countries both because of their unilateral option and because of their generally lower level of enthusiasm for international organization. The exercise of that option, however, seemed likely to entail high costs in other law of the sea areas in which some sort of international agreement was seen as necessary (straits, economic zone). In this sense the unilateral option of the advanced technology states on the deep seabed issue was largely offset by the unilateral options of the developing states on coastal issues. Furthermore, nearly all of the advanced technology states accepted the 1970 declaration of principles and thus the idea of an international solution based on the common heritage principle. This acceptance made it difficult, though by no means impossible, for these states to revert to unilateral solutions, and established the declaration of principles as a major landmark in the stability of the deep seabed negotiations. Up to the autumn of 1975, these factors had firmly stemmed any impulse to unilateralism, and seemed likely to continue to do so as long as the international negotiations maintained an acceptable momentum toward agreement within an acceptable time frame.

In this sense, the outcome of the Geneva session might have reinforced the commitment of the advanced technology states to the negotiations because of the contents of the single text (for C.I) but weakened it because of the length of time remaining before a convention was finally produced. Should the next session of the negotiations stall for any reason—for example, the problems for the Group of 77 in finding a response to the single text—then time might take precedence over substance in shaping the attitude of the advanced technology states. Similarly, should the developing coastal states force too harsh a solution on coastal zone issues, then incentives for restraint on the deep seabed by the advanced technology states would decline.

In sum, then, the commitment to the international level of action of one, and possibly both, of the two important groups of states involved on the deep seabed issue is precarious. It is sensitive not only to the substance and the time frame of the deep seabed negotiations, but also to other issues.

Substance

Analysis of the factors shaping the substance of the deep seabed negotiations, as noted above, is even more difficult than for the continental shelf, principally because the nature of the outcome is both less clear and less certain to be implemented. Not only is the single text from the Geneva session unlikely to survive without very substantial revisions, but also the maintenance of states' commitment to extended negotiations is uncertain. Nevertheless, some observations are possible.

An important background factor has been the influence of the freedom of the seas idea left over from the 1958 Convention on the High Seas. The widespread general acceptance of this idea, and the international rather than national status for the deep seabed that it included, provided fertile ground for Pardo's proposal in a way that the Continental Shelf Convention clearly did not. It also provided a fallback position for the advanced technology states in the event of their not accepting an international regime (that is, the deep seabed as an unsupervised, unclaimable, common area rather than as some form of supervised common property).

Given this basic framework, Pardo's proposal decisively shaped virtually every other substantive aspect of the issue. The full extent of his initiative's influence is perhaps best illustrated by the declaration of principles, which, although it was separated from his proposal by three years of negotiation in the SBC, essentially mirrored its major provisions. Even after eight years of negotiations, the substance of the debate in C.I closely reflected Pardo's original ideas. If the negotiations come to almost any sort of successful conclusion, then the influence of Pardo's key ideas like the common heritage and the international regime and machinery will have been enormous.

This prospect is, however, threatened in two ways by factors related to the time element. First, the prospect of any international settlement is being reduced by the increasing chance of unilateralism by the advanced technology states. Second, the mere passage of time allows intervention by other factors that weaken the momentum of the initiative either by eroding its substance or by lengthening the time required to implement it, and so increasing the chances of the first threat materializing. Examples of intervention by other factors include the settlement of the continental margin boundary in C.II in such a way as to give the least area and fewest resources to the international area (though this was arguably inevitable given the existing strength of the coastal state claims and the inextricable link between the margin and deep seabed boundary questions) and the development of a unilateral option by the advanced technology states that has strengthened their position in the negotiations and thereby involved compromises away from the direction of Pardo's original proposal. Examples of additional slowing factors are the raising of the whole law of the sea issue (particularly its coastal zone aspects, which not only drained energy and attention away from the deep seabed issue but also complicated both its substance and the politics surrounding it) and the evolution of the global debate on commodities and the new economic order, which have added to the already heavy burden of politicization that the international seabed regime issue has had to bear. These problems of process resulted partly from the character of Pardo's proposal itself, and to some extent Pardo foresaw their danger. To a considerable extent, however, they were inherent in the international negotiating process, a difficulty to be discussed in more detail below.

The political factors shaping the outcome of the deep seabed issue have been quite different from those shaping the outcome on the continental margin. In the case of the margin, the decisive factor appears to have been the overwhelming strength of one side in relation to the other. In the case of the deep seabed, the decisive factor has been the relatively equal strength of the two sides, compounded by their relatively inflexible negotiating positions and the consequent obligation to compromise that this imposes on both of them.[2] This has defined a long and laborious process of negotiating in an effort to reconcile the developing countries (inspired by the Pardo proposal and the claims of the new economic order) with the economic concerns (access, security, and profit) of the advanced technology states. The economic and technological unknowns of the issue have contributed to the further politicization of information, which has in turn complicated and slowed the negotiations even more. The whole delicately balanced situation has opened up interesting roles for mediating individuals like C. W. Pinto, whose tireless efforts in search of workable compromises have made him an important influence on the outcome.

Any final assessment of the shaping factors on the deep seabed issue must await the materialization of a conclusive outcome—an event that seems likely to be some years in the future. Two general types of outcomes are possible: the breakdown of the negotiations either followed or preceded by unilateral claims on the part of one or more advanced technology states, or the conclusion of the negotiations by some sort of agreement on an international regime and machinery.

In the case of a breakdown, the two main shaping factors would be the power of unilateral action given to the advanced technology states by their command of the appropriate technology, and the whole array of factors working to slow down the negotiations sufficiently to give that option time to come into play. Even in the event of an outcome along these lines, however, it seems likely that some influence from Pardo's initiative would survive. The United States, for example, has accepted the idea of the common heritage, and a government version of the Hard Minerals Bill would almost certainly contain at least some provisions for revenue sharing with developing countries. Should events go this way, one likely effect would be the undermining of the agreement on continental shelf limits. Developing coastal states, probably led by the Latin Americans, might seek revenge by staking claims to huge areas of deep seabed beyond the margin, possibly reviving the long discredited "national lakes" approach to the seabed, if not the oceans.

In the case of a negotiated agreement, the influence of Pardo's initiative as a shaping factor would be much greater—how much greater would depend on the influence of extended time on the relative strengths of the two sides in the negotiations. In different ways, the passage of time strengthens both sides, and this in itself helps explain the weakness of the attempts to fix deadlines to the process. The advanced technology states benefit from the steady maturation of their unilateral option, but that option is also, as discussed above, dangerous to the negotiations. The developing countries benefit inasmuch as time is on their side in terms of the new economic order movement. That this movement is making palpable changes in the attitudes and behaviors of the industrialized states is clear from events such as the Lome Convention between the EEC and a large group of developing countries, and the conciliatory attitude taken by the United States on global economic matters at the UN's seventh special session in September 1975. Thus, from the Group of 77's point of view, the longer the negotiations run on the more likely are the advanced technology states to move toward the Group's position. In addition, the developing countries also benefit from the increasing capabilities of the Group of 77 as an organization, in terms of more efficient structures, better command over expert advice and information, and more experience as an actor in world politics. The remarkable evolution of the Group since its early days has been particularly noticeable in the debate on the new economic order.

At this point one can only guess at which side is likely to benefit most from the plowman's pace of the negotiations. The main point is that neither side would benefit from the collapse of the negotiations, an outcome that looms large beneath their efforts to maneuver for advantage.

CONCLUSIONS

This examination of the interaction model for seabed politics, and also of the three basic historical/political questions relating to the subject, leads to conclusions in three areas.

The first area is the strengths and weaknesses of the various alignments of states that emerged as actors in seabed politics. Although no one type of alignment was conspicuously successful compared to other types, some observations can be made about the factors that strengthened or weakened alignments of all types. Much of this analysis has already been carried out in previous chapters, and it is not necessary at this point to review the whole list of strengthening and weakening factors. Four observations, however, are worth emphasizing.

First is the way in which the power of groups based on large blocs of votes is diminished by consensus-oriented rules of procedure. This seems to be the price that weak majorities will have to pay in order to induce strong minorities to participate in any such negotiations in the future. Its effects are to enhance the influence of other factors and to slow down the process of negotiation.

Second is the way in which time factors affect the strength of some alignments. In the law of the sea negotiations, for example, the passage of time strengthened the positions of the coastal states in general, the territorialist group, and the advanced technology states, and weakened the position of the landlocked and geographically disadvantaged states and the maritime states. This influence is

FIGURE 11.3

Summary of Key Factors Shaping Negotiations on the International Seabed Regime and Machinery

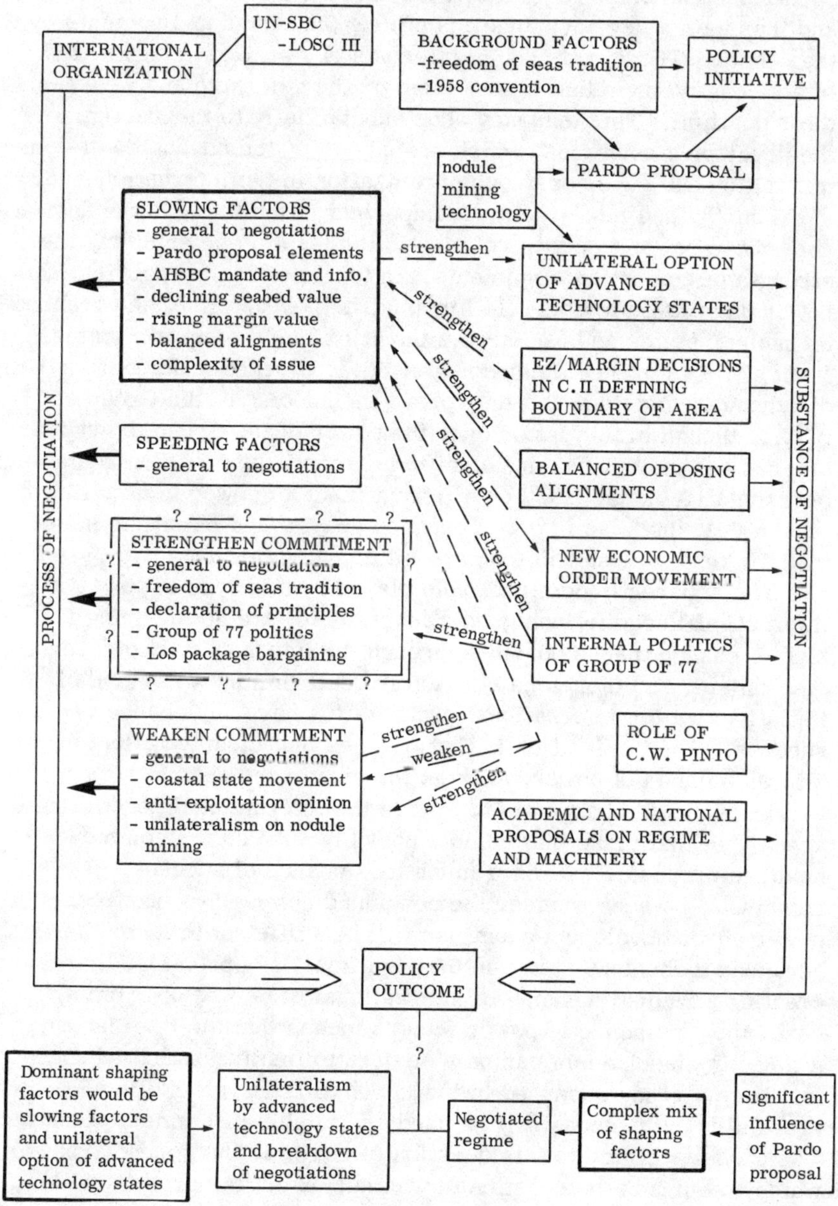

Source: Compiled by the author.

particularly important in the light of the large role played by time factors in the international negotiating process generally.

Third is the role played by attitudes arising from the pattern of vested interests in the issue at hand. Those with little to lose and much to gain, in this case the coastal states and the developing countries, can seize the initiative in the negotiations rather easily and thus gain a key advantage in defining the terms and contents of the debate. Those with vested interests, by contrast, tend to fall into a reactive role that places them on the defensive and weakens their position. This tendency adds substantially to the strength of the developing countries, which already enjoy the advantage of voting majorities and superior group organization in the international forum.

Fourth, and much the most important of the four, is the decisive role of the power to act in relation to the issue in determining the strength of competing alignments. In terms of seabed politics, this power has played a vital role for small minorities like the advanced technology states and for large minorities like the margin states. Lack of it has proved the undoing of the landlocked and geographically disadvantaged states, and may prove the undoing of the Group of 77 in C.I, though in this latter case the power of the Group to act in other areas may prove sufficient to hold the balance. Confrontations between alignments based on different kinds of power to act, like that in C.I, and also like that between coastal and maritime states in C.II, tend to result in long and bitter struggles.

The apparently dominant role played by this type of power in international negotiations goes a long way toward determining the kind of change that is possible through the international conference mechanism, and therefore also toward determining what kind of initiatives within international organization have any chance of success. The implication is that changes that clearly reflect an altered balance of power, such as the rise of the coastal states, will have a good chance of success in the international negotiating arena, but that those that do not, possibly such as the common heritage proposal, will have much less chance of success. It remains to be seen whether the new found commodity-producer power of the developing countries will be sufficient to lever the advanced technology states into a significant compromise, thereby creating a radical advance in international organization. In any case, the dominance of power factors means that the international negotiating mechanism can only be used to institutionalize changes that have already occurred in the international system. In all probability, it cannot itself be used as a major instrument of change.

Closely related to this question of power is the second area, which comprises the question of whether or not the process within international organization leading from initiative to policy outcome is workable, particularly in respect to innovative expansions of the structures, powers, and functions of international organization. Since the negotiations have stretched on much longer than originally anticipated, the full evidence on this question, in the form of final success or failure of the negotiations, is not yet in. On the basis of the negotiations to date, some writers are already beginning to argue that the process is not workable.[3]

The crucial variable in this argument is time. If the negotiations proceed too slowly, then opportunity is created for all manner of factors to influence them. These factors include technological and economic developments (the nodule mining industry), political developments (the new economic order movement), developments in related issues (the economic zone), and developments caused by feedback from the international level of action to other levels (national claims). They appear to have four kinds of effect on international negotiations. First, they can make them more complicated in terms of both issues and alignments, and thereby slow them down even further. Second, they can weaken the shaping influence of the original initiative either by slowing its implementation or by undermining its substantive aims. Third, they can weaken the commitment of states to the international negotiations. And fourth, they can change the balance of power between contending alignments and thereby alter the substantive outcome.

If the negotiations are sufficiently slow, then at some point the additional slowing and disrupting impact of these factors will cause the whole international process to collapse, taking with it most of the shaping force of the original initiative and moving the political process back to unilateral, regional, or functional subglobal levels of action. Alternatively, if the major impact comes in the form of changes in the power balance between contending alignments, then the international initiative might be overriden by national initiatives and the international process turned to codification of state practice.

No such collapse or alteration has as yet overtaken the international process of seabed politics, though both alternatives are very likely outcomes. Indeed, on the basis of the work in C.I up to this point, one or the other of these alternatives looks like a more probable outcome than any settlement along the lines of Pardo's proposal. The weight of evidence from seabed politics thus seems to favor a negative answer to the question of whether the international process is workable in terms of major advances in international organization. For the seabed at least, states appear to have been able to stop or to outrun the international initiative toward radical developments in international organization before those developments could be institutionalized and secured. This outcome points to the conclusion that international organization as currently structured cannot, even when given the advantage of a strong initiative, produce more than incremental changes in the state-dominated international system. If seabed politics is a valid example, then those in search of means to radical reform of the international system should conduct their search elsewhere than in the corridors of the United Nations.

But is seabed politics a valid example? Are the forces that have slowed and crippled the attempt to institutionalize the radical vision of the common heritage of mankind an inevitable part of the international process?

Many of the structural and process factors identified in this discussion as having a slowing effect do appear to be fixed features of the international negotiating process on any controversial global issue. These include the number of states participating, the character of major actors (the United States, the Group of 77), disruptive actions by states, the existence of ideologically defined opposed positions, the politicization of information, the ignorance of many delegates, and the tendency for policy alternatives to proliferate. In addition, because the international process gives power (votes) to states that would otherwise be weak, slow and protective procedures like those adopted in the law of the sea negotiations also seem inevitable.

Some, however, appear to be at least partly avoidable by more aware planning and organization of any international action. These include the vague mandate given to the AHSBC, the inadequate preparation for the Conference, the length and complexity of the agenda, the matter of deadlines, and the design of the original initiative. Unfortunately, all but the last of these are subject to manipulation by states within the international process, so that even if preparation was more careful than in the current case, the actions of vested interests could make the outcome much the same.

The design of the initiative itself is thus the point on which this question hinges. The profound shaping influence that such an initiative has may give it sufficient weight to tip the issue one way or the other. Pardo's proposal has been criticized here for introducing elements of conflict into the negotiations that may not have been necessary to the achievement of its goals, and that may have achieved ideological purity at the expense of concrete influence on the outcome of events. Whether or not the influence of these elements was in fact decisive in the success or failure of Pardo's initiative is a question that cannot be answered conclusively. The lesson can nonetheless be taken that the design of such initiatives in the future should be approached with extreme care and with full awareness of the massive multiplier effects that even a small verbal input can have when properly placed in the high energy system of international politics. Other than an unexpectedly successful conclusion to the negotiations in C.I, this single factor is the only variable that might invalidate the pessimistic conclusions reached above about the practicability of initiatives aimed at radical changes in the state-dominated international system.

The third area is the matter of what implications the political process revealed in this study has for the international law that is its outcome. Four observations are salient here.

First, it is important to emphasize that international law-making by the conference method is primarily a political process. Its sources are as diverse as national claims and Pardo-type initiatives, and it is subject to change as the political configurations on which it is based change. In this sense, international law can be seen as an agreed set of rules among states (and other entities) with the power to act in relation to given issues. The rise of the coastal states, and also of the developing countries, illustrates the process by which international law changes as new actors come onto the stage.

Second, the political process of drafting international law leaves signposts in the final texts to mark areas where no concrete agreement could be reached. These signposts take the form of vaguely worded compromises, and are often the focus of difficulties in practical implementation of the law. Examples of this are the exploitation limit for the continental shelf and the peaceful purposes principle in Resolution 2749. Assuming a successful outcome in C.I, places where such signposts are likely to occur are in the texts covering the powers and functions of the authority, the conditions of exploitation, and the mechanisms for compensating land-based producers. Vague compromises here could create severe difficulties in the complex task of implementing new laws in relation to the common heritage. Among these difficulties could be the tendency for states to seize upon weak points in international law in order to mount major campaigns for revision, as was done most notably by the Latin American states in relation to the continental shelf and the territorial sea.

Third, the underlying commitment of most states to international law is surprisingly strong. This is indicated not only by the amount of effort states have invested in the negotiations, but also by the degree to which many have been prepared to compromise. It is also indicated, in a paradoxical way, by the effectiveness of threats to wreck the negotiations and by the restraint of states in implementing such threats. Thus, what often appears to be the imminent collapse of international negotiations is frequently just a noise-making process that is part of bargaining tactics. If nothing else, the law of the sea negotiations have demonstrated that such negotiations are often much less fragile than they appear to be, though this conclusion should in no way be taken to suggest that an increase in pressure on such a conference would not cause its breakdown.

Fourth, regardless of whether or not the Law of the Sea Conference succeeds in producing a convention, its proceedings are going to have an enormous effect on international law. Even in the case of a complete breakdown of the conference, subsequent national actions would largely follow the outlines of the major compromises reached in the negotiations. While this would not prevent the conflicts that caused the breakdown, it would create a much greater degree of stability and uniformity in customary law than would have occurred through a random process of national claims. Thus the conference might substantially succeed in practice (in terms of shaping new international law) while failing in appearance (in terms of producing a convention). This possibility makes the whole process look much more worthwhile than if success or failure were measured only by the ability to produce a convention.

NOTES

1. For a more detailed analysis of these and other slowing factors, see Edward Miles in Francis T. Christy et al. (eds.), Law of the Sea: Caracas and Beyond (Cambridge: Ballinger, 1975), pp. 39-47, in which the argument is made that the UN should never again convene a conference of the size and complexity of the Third Law of the Sea Conference because of the inevitable impact of the built-in slowing factors.

2. The inflexibility of the Group of 77 arises largely out of the extensive internal bargaining and compromising necessary to arrive at a common position among its more than 100 members; that of the United States arises largely out of the extensive internal bargaining and compromising necessary to arrive at a common position among its numerous government bodies, agencies and departments, and influential private pressure groups. (See Ch.8, n. 40.) A comparison of the two as "dinosaurs" at international conferences would be most interesting.

3. Miles and Douglas Johnston, Marine Policy and the Coastal Community: Studies in the Social Sciences (London: Croom Helm, 1976) pp. 39-47, 357-72, 400-01.

POSTSCRIPT

In terms of seabed politics, the New York session (March 15- May 7, 1976) brought substantial refinement of the already stable compromise on the continental shelf, and a major confrontation in Committee I on the international seabed regime and machinery.

The New York session saw a considerable blurring of the differences between the flat 200-mile and the 200-mile-plus margin schools. Tensions surrounding the issue of the margin beyond 200 miles declined noticeably, at least in part because the landlocked and geographically disadvantaged states pushed their demands for rights on the shelf less stridently. The margin states maintained their compromise offer to share revenues from the margin beyond 200 miles, though Australia and Argentina still held back on this, and the United States tabled a firm offer on the modalities of such sharing. This involved a five-year grace period during which no sharing would be required, followed by five years during which sharing would rise by 1 percent per year until it reached 5 percent of production value. After that, it would remain stable at 5 percent per year for the duration of production. Although this offer was not formally discussed, it appeared to be broadly acceptable to the other margin states that accepted the principle of sharing.

The margin states also succeeded in reaching agreement on a definition for the edge of the margin. This was based on a much improved version of their Geneva effort, and involved a formula for determining the precise limits of coastal state jurisdiction on the rise. According to this formula the coastal state could only claim areas of the rise where the thickness of the sediment was not less than 1 percent of the distance to the base of the slope (that is, a point at which the sediment of the rise was two kilometers thick could only be claimed if it was within 200 kilometers of the base of the slope). The effect of this, if implemented, would be to give margin states the most valuable areas of the rise. It would, however, lay to rest fears that margin states might pursue creeping jurisdiction out to the last grain of continental sediment.

By contrast, the tradeoff on rights of access for non-coastal states to fisheries in the economic zone became a focus of contention. Non-coastal states were unhappy with the provisions in the single negotiating text (SNT), and coastal states tended to resist discussion of

---

This book was going to press as the New York session was drawing to a close. Consequently, this postscript was written on the last day of the session and is an initial impression of events. No attempt has been made to analyze the revised single negotiating texts in any detail, but it is hoped that the observations offered herein will be useful in clarifying some of the discussion in the preceding two chapters.

specific sharing arrangements.

The revised SNTs, which once again did not appear until the last day of the session, did not contain the precise definition of the edge of the margin, and included only minor alterations on the rights of non-coastal states to fish in the economic zones of neighbors. The outlines of the United States' sharing offer on the margin beyond 200 miles was included, but without the figures on percentages.

In Committee I, the situation was much more difficult to assess. For most of the session, C.I. functioned with remarkable smoothness. There was no outcry over the Geneva SNT, and delegates spent nearly the whole session working on what appeared to be a revised text containing many very substantial compromises. This dramatic shift to a relatively conciliatory atmosphere could be explained by three factors. First, the ideological leadership of Algeria in the Group of 77 appeared much weakened. Second, the threat of unilateral action on nodule mining by the United States loomed much larger than at previous sessions. And third, the Group of 77 appeared to have broken ranks, with a significant group of Latin American and Asian states, including several land-based metal producers, prepared to negotiate a deal with the advanced technology states.

By the end of the seventh week of the session, a nearly complete revised SNT had been drawn up, largely on the basis of negotiations between the advanced technology states and the moderate members of the Group of 77. This text was, on the whole, favorable to the advance technology states, and the problem remained of whether or not the moderates could persuade the rest of the Group of 77 to accept it. The Gro of 77, particularly the African states, refused to accept the new text as a replacement for the one worked out at Geneva. Their refusal once ag illustrated the symbolic strength of Group of 77 unity, and precipitated an acute crisis during the last few days of the session.

This crisis involved two issues: first, what form should the revised SNT for C.I. take; and second, what date should be set for the next session of the conference? Both issues were critical in that a failure on either would greatly increase the pressure for, and liklihood of, unilateral action on nodule mining by the United States. The United States de gation made abundantly clear the volatile condition of American politics on this issue, in effect threatening a breakdown of the conference unless both issues were resolved in favor of American interests. At a minimu this appeared to involve a further session of the conference this summe and an SNT for C.I. that was not too far out of line with that drawn up a the session. The United States won the day on both issues after exertin an enormous amount of pressure. A threat by the Group of 77 to delay the next session until early 1977 was averted, and the conference agree to meet again in New York from August 2 to September 17.

As it did at Geneva, the conference has again ended on an ambiguous note. Its future seems assured for at least another session, but its slo progress continues to run a very close race with the numerous factors tending to undermine it.

INDEX

ABOUT THE AUTHOR

BARRY BUZAN is a research fellow at the Institute of International Relations, University of British Columbia, Vancouver. He attended the 1974, 1975, and 1976 sessions of the Law of the Sea Conference, and has published widely on the subject during those years. His articles have appeared in The Canadian Yearbook of International Law, International Perspectives, and the Law of the Sea Institute's Occasional Paper series, as well as in a number of newspapers and specialist journals.

Dr. Buzan earned his Ph.D. at the London School of Economics.

THE INTERNATIONAL POLITICS OF MARINE POLLUTION CONTROL
Robert A. Shinn

OCEAN SPACE RIGHTS: Developing U.S. Policy
Lawrence Juda

INTERNATIONAL ENVIRONMENTAL LAW
edited by Ludwik A. Teclaff and Albert E. Utton

THE WORLD FOOD CONFERENCE AND GLOBAL PROBLEM SOLVING
Thomas G. Weiss and Robert S. Jordan

PLANNING ALTERNATIVE WORLD FUTURES: Values, Methods, and Models
edited by Louis Rene Beres and Harry R. Targ